W9-BUZ-482

SNAKES OF THE SOUTHEAST

Snakes

OF THE SOUTHEAST

by Whit Gibbons and Mike Dorcas

The University of Georgia Press
Athens and London

Athens, Georgia 30602

Designed by Mindy Basinger Hill

Set in 10/15 Scala

Printed and bound by Four Colour Imports

The paper in this book meets the guidelines for permanence and durability of the Committee on Production Guidelines for Book Longevity of the Council on Library Resources.

Printed in China

09 08 07 06 05 P 5 4 3 2 1

Library of Congress Cataloging-in-Publication Data

Gibbons, J. Whitfield, 1939–

Snakes of the Southeast / by Whit Gibbons and Mike Dorcas.

p. cm. — (A Wormsloe Foundation nature book)

Includes bibliographical references (p.) and index.

ISBN 0-8203-2652-6 (pbk. : alk. paper)

1. Snakes—Southern States. I. Dorcas, Michael E., 1963–
II. Title. III. Series.

QL666.O6G35 2005

597.96'0975—dc22 2004021098

British Library Cataloging-in-Publication Data available

Contents

SNAKES OF THE SOUTHEAST

Rainbow snakes are among the most colorful of the southeastern snakes.

Some snakes, such as this cross between a red milksnake and Louisiana milksnake, can be strikingly colored.

All about Snakes

WHY SNAKES?

Practically everyone is fascinated by snakes. Some people fear them, some are attracted to them, but almost nobody is indifferent toward them. And although dread of snakes may be the most common phobia in the Southeast, today's soaring interest in environmental issues has led people to pay more attention to the environmental well-being of all wildlife, including snakes.

Most people are interested in learning which snakes live around their homes, which are venomous and which are harmless, and how to tell them apart. Our goal is to teach people about snakes and to foster appreciation of them as valuable components of our natural heritage. Snakes have been feared, hated, and maltreated for too long, and have been too long ignored when conservation and environmental issues are under discussion. With this book we hope to interest young people and

Many species of snakes, including the eastern hognose, do not bite people.

Black swamp snakes are rare in most regions but can be very abundant in localized areas.

A harmless red-bellied watersnake

adults who may have missed an earlier opportunity to get to know this group of captivating yet harassed animals.

The Southeast offers many opportunities for appreciating nature, and our hope is that people who read this book will become eager to see snakes when they enter the woods, boat on a lake, or walk along a stream. We encourage people to place the same value on encountering a snake as they do on seeing a dolphin, hearing a screech owl, or touching a box turtle. We would like to see everyone develop an acceptance of—better yet, an admiration for—snakes that equals that expressed for many other wild creatures.

Some snakes, such as the eastern hognose snake, can sometimes be found in suburban areas.

GENERAL BIOLOGY OF SNAKES

What is a snake? Snakes are reptiles, just like lizards (their closest relatives), turtles, and alligators. Their most distinctive traits are their elongated body and lack of limbs. These characters impose certain limitations on snakes compared with other animals, yet these same biological features give them unusual abilities as well. Snakes can maneuver through underground burrows and tunnels and negotiate tight passages much better than most other animals of their size. This ability, added to their agility, helps them find prey and hide from predators.

Legless lizards, such as this slender glass lizard (*Ophisaurus attenuatus*), are frequently misidentified as snakes. Glass lizards have movable eyelids and ear openings whereas snakes do not.

Like most other reptiles, snakes are covered with scales. Even their eyes are protected by clear, transparent scales, which eliminate the need for movable eyelids. The shape, size, and placement of the body scales in relation to each other are different for each species and are commonly used in identification. All snakes shed the outermost layer of skin covering their scales several times per year. Most snakes shed by literally crawling out of the skin, leaving behind an inside-out remnant. Scientists call this shedding process *ecdysis*.

Did you know?

Some southeastern lizards have no legs and are often mistaken for snakes.

All snakes have teeth. Some, including six species found in the Southeast, have hollow fangs in the front of the mouth that are used to inject venom. Some snakes are rear-fanged, which means they have enlarged teeth in the back of the mouth. These snakes are generally harmless to humans. Most snakes have many thin, needlelike teeth in their upper and lower jaws that curve backward, so that a captured prey animal cannot escape. Snakes swallow their prey whole, and they can eat animals much wider than they are themselves because their upper and lower jaws are loosely connected to each other and their bodies stretch. Snakes may take more than an hour to swallow very large prey.

The relatively narrow, elongated bodies of snakes require an internal anatomy that differs somewhat from that of other vertebrates, including other reptiles. Whereas paired organs are usually side-by-side in most animals, they tend to be staggered in snakes. For example, one lung is farther down the body than the other, and in most species only one lung actually functions. Likewise, the kidneys are placed one ahead of the other, as are the testes in males and the ovaries in females. The liver is greatly elongated compared with that of a mammal, bird, or reptile of similar size.

The digestive tract of a snake is like those of other vertebrates in having an esophagus, stomach, and small and large intestines that lead to the outside through the *cloaca*, a chamber into which the intestinal, urinary, and genital tracts empty. The cloacal opening, visible on the underside of all snakes, is where the body ends and the tail begins.

The eye becomes opaque shortly before a snake sheds its skin, as seen in this eastern coachwhip.

The reproductive organs of snakes are also of the same basic design and function as those of other vertebrates, except that male snakes have two copulatory organs (i.e., penises) called *hemipenes* (singular = *hemipenis*). Unless the snake is copulating, the two hemipenes are tucked inside the base of the tail behind the cloaca. Thus, male snakes typically have longer and thicker tails than female snakes. Dur-

Approximately half of the species of southeastern snakes lay eggs.

ing mating, the male snake fertilizes the female by inserting either hemipenis into her cloaca. Sperm travels along grooves in the hemipenis and into the female to fertilize her eggs. Some female snakes can store sperm for months or even years.

Snakes have the same senses as other vertebrates. Although they cannot hear airborne sounds the way mammals do because they have no ear openings and no middle ear, snakes can detect vibrations through

the ground or water. Their combined sense of taste and smell is well developed. The snake flicks its forked tongue to gather odor particles from the environment. The tongue transports the particles to cavities in the roof of the mouth called *Jacobson's organs* for chemical identification. Snakes that typically travel above the ground have very good vision and often have enlarged eyes, whereas some burrowing forms have reduced eyes and can discern only shadows. Because snakes have no eyelids, they do not blink and seem always to be staring.

Some snakes, including the southeastern pit vipers, have special sense organs in the head region that can detect variations in temperature. The heat-sensitive organs of pit vipers are in openings called "pits" on the face of the snake between the eye and the nostril. Even in complete darkness a pit viper can detect and pinpoint a subtle change in temperature such as the difference between the body temperature of a mouse and its surroundings. A sense organ that can detect heat has great value to an animal that hunts at night.

Facing page and above:
Timber rattlesnake shedding its skin. The clear scales that cover the eyes are shed with the rest of the skin.

Did you know?

Snakes do not hypnotize other animals. Some snakes move very slowly toward their prey, and some actually move the head slowly from side to side as they approach, which may resemble a twig blowing in the wind. This may help the snake get close enough to capture its prey. Sometimes the prey animal does not move because it does not realize the snake is there; or it may be waiting for the right moment to attempt to escape.

SNAKE DIVERSITY

Although all snakes are superficially alike in lacking limbs, the diversity of body forms, habitats occupied, feeding strategies, and behavior patterns they display is remarkable. Approximately 15–18 families of snakes comprising more than 2,800 species are known worldwide. More than 130 species in 5 families are native to the United States. The Southeast is the natural home to more than 50 species belonging to 3 families: the Elapidae (the coral snake), the Viperidae (cottonmouth, copperhead, and rattlesnakes), and the Colubridae (all remaining native snakes). Two introduced species—the Brahminy blind snake and the Burmese python—have become established in some areas of Florida, adding 2 more species and families to the total. About a dozen of the species native to the Southeast are *endemic*, which means that they are found nowhere else in the world.

Snakes of the Southeast belong to one of three families: Colubridae (top), Viperidae (middle), or Elapidae (bottom).

Southeastern snakes occupy, or at least enter, virtually every natural habitat. Sixteen species, including the venomous cottonmouth, rely on aquatic habitats for their primary prey of frogs, fish, or aquatic invertebrates such as crayfish. Some species are characteristically found in sandhill habitats; others are most likely to be found in hardwood or pine forests, vegetation along stream margins, rocky outcrops, or open fields. The salt marsh snake is the only species of North American snake that lives permanently in brackish water.

Naming Snakes

All animals are given a scientific name that includes a *genus* name (e.g., *Coluber*) and a *species* name (e.g., *constrictor*). Thus, the proper scientific name of the racer is *Coluber constrictor*. Scientific names follow strict rules. Common names do not follow the established rules and are often based on what the people of a particular region call the snake. For example, the snakes known scientifically as *Elaphe obsoleta* are called "rat snakes" by most herpetologists (scientists who study snakes, other reptiles, and amphibians; see page 221), "black snakes" in northern areas, and "chicken snakes" in many parts of the South. We see no compelling reason to insist on a strict use of a common name for any species of snake as long as it is clear what species of snake is being referred to, especially if the customary terminology varies from one region to another. In addition, forced standardization of common names often detracts from the regional cultural heritage and is unnecessary for professional herpetol-

ogists who use the scientific names. In this book we use scientific names to minimize confusion plus common names that are recognized by most people in the Southeast.

Watersnakes are common residents in all natural aquatic areas in the Southeast.

A Word about Taxonomic Controversies

Taxonomy is the scientific field of classification and naming of organisms. Snake taxonomists strive to classify and name snakes in a way that reflects the ancestral relationships among species. Thus, closely related species are placed together within a genus (plural = *genera*), and closely related genera are grouped together within a family. Taxonomy is not a perfect science. Very rare snakes may be difficult to classify because so little is known of their biology. In addition, taxonomists often disagree about the relative importance of different traits in classification, and thus may have different views on the lineage

Did you know?

Of the 53 species of snakes native to the Southeast, 24 are found in every one of the southeastern states.

Salt marsh snakes live in a variety of brackish water habitats.

Ribbon snakes are often found on land near water.

Crowned snakes are among the smallest southeastern snakes.

and ancestral relationships (*phylogeny*) of particular species or groups of species. Sometimes these disagreements result in the scientific name of a species being changed. Modern molecular genetics has allowed herpetologists to resolve many of the old taxonomic debates, although in some instances disputes have actually become more heated because of differing interpretations of the findings.

In this book we have used the scientific names familiar to most herpetologists. We have included a short explanatory note for each species whose classification is currently in dispute. Our purpose is to be certain that the reader knows what snake we are talking about, no matter what name the animal is given.

FOOD AND FEEDING

Did you know?

Approximately 140 distinct species of snakes are found in the United States and Canada. If subspecies are included, the number is more than 300.

All snakes are carnivores (that is, they eat animals rather than plants), and all snakes swallow their prey whole. Although some species have very specialized diets, snakes as a group eat almost every kind of animal. Small snakes eat insects, spiders, and earthworms; larger ones eat rabbits, birds, and bullfrogs. Overpowering and consuming other animals despite having no limbs is a remarkable feat.

The first step in getting food is finding it. Among the senses used by southeastern snakes to find and identify their prey are sight, smell, touch, and infrared heat detection. Some species, such as racers and coachwhips, use their particularly good vision to find prey. Many snakes track prey by following its scent. Pit vipers, such as rattlesnakes, use scent to arrive at appropriate ambush locations and then use their infrared detection

Rattlesnakes rely on camouflage to ambush prey.

pits to home in on the body heat of mammalian prey, even in the dark. Rattlesnakes usually strike, inject venom, and release the prey, then rely on their remarkable ability to track down the dying animal after it runs away. A brown watersnake will drape itself over a limb, hang its head and upper body into the water, and grab, with spectacular speed, any fish that touches it.

Herpetologists categorize snakes into two broad categories based on how they hunt their prey. Ambush predators, such as diamondback rattlesnakes, may remain motionless for hours or even days waiting for prey to pass within striking distance. Juveniles of many species of pit vipers, including copperheads, cottonmouths, and pigmy rattlesnakes, wiggle a brightly colored tail tip to attract potential prey, such as frogs or lizards, within striking distance. Wide-ranging, active foragers, such as racers and coachwhips, cover large areas in search of prey. Some snakes use both strategies, ambush and active foraging, to find prey.

Many snakes track prey by using their tongues to follow a scent trail.

Snakes can also be categorized by the type of food they eat. *Generalists* eat a wide variety of animals, including recently dead ones; *specialists* focus on one or a few prey types. Racers and kingsnakes are good examples of

Some snakes specialize on particular prey items, including (top to bottom) scarlet snakes on reptile eggs, coral snakes on other snakes, some watersnakes on fish, and some crowned snakes on centipedes.

generalists because they eat many types of animals—including birds, mammals, reptiles, and even other snakes. Hognose snakes, which eat primarily toads, are dietary specialists.

Snakes have several techniques for subduing prey. The most straightforward approach is to bite the animal to be eaten and hold on while gradually swallowing it. All southeastern snakes have numerous sharp, backward-pointing teeth that direct the prey in one direction—toward the stomach. The indigo snake has a powerful bite that crushes soft-bodied prey. Constriction is another method widely used to kill living prey. After striking, the snake loops its body once or more around the prey and tightens the coils each time the prey animal exhales. Some herpetologists think that suffocation is the primary cause of death; others believe that constriction results in cardiac arrest. Quite a few snakes—more than you might think—use venom to subdue or kill their prey. The saliva of many species can paralyze or even kill small animals when the venom enters wounds made by the sharp teeth. But only six southeastern species—the five pit vipers and the coral snake—actually inject venom.

Some of the specialist feeders have unusual structural, physiological, and behavioral strategies that allow them to capture and eat their chosen prey. Crowned snakes eat centipedes, which are venomous. The crowned snake bites the centipede behind the head in a way that prevents the centipede from biting back. The snake forces its enlarged rear teeth into the centipede's body and holds on tightly until its venomous saliva has paralyzed the centipede. Mud snakes and rainbow snakes have pointed scales on their tail tips that they use to secure themselves while eating giant salamanders or eels. The glossy crayfish snake and striped crayfish snake use their enlarged, chisel-like teeth to grasp and consume hard-shelled crayfish.

Scarlet kingsnakes kill their prey with constriction.

Did you know?

The primary diet of some snakes is other snakes.

Snakes often eat prey items larger than their own girth.

Virtually all snakes can go long periods without eating—many days, weeks, or in some instances months—and in cooler parts of the Southeast, most do not eat at all during the winter. Likewise, by remaining dormant underground, most snakes are able to survive periods of drought when food may be scarce.

PREDATORS

Snakes and snake eggs are eaten by an enormous variety of animals throughout the Southeast. Almost any large animal that eats other animals will eat a snake if it can.

The most vulnerable snakes are the small ones, which must fear small creatures such as spiders, toads, and shrews in addition to larger predators. Large carnivorous mammals such as bobcats and coyotes, as well as medium-sized ones such as raccoons, skunks, and opossums, are potential threats to larger snakes, although the snake's size can become a deterrent in some instances. Otter and mink readily prey on snakes in wetland habitats. Birds of prey, especially hawks, prey on snakes they find crossing open areas and lying on limbs along riverbanks. Snakes that travel overland at night fall victim to owls; small woodland snakes

Domestic cats kill many small snakes in suburban areas.

Raccoons and bobcats are natural predators of many snakes.

Any snake, including venomous pit vipers such as this copperhead, may become the prey of a common kingsnake.

American alligators may eat any snake that enters their aquatic habitat.

become prey for ground-scratching birds; and aquatic snakes along wetland margins become the quarry of wading birds. Domestic and feral cats kill thousands of snakes each year in residential areas.

Large fish such as gar, catfish, and bass are an ever-present hazard for aquatic snakes, and alligators are a constant menace to all snakes in the water, even large cottonmouths. Some amphibians eat snakes as well. Bullfrogs and the giant salamanders known as amphiumas are common predators of small to medium-sized aquatic snakes.

Although many predators will eat snakes when given the opportunity, only one type of predator in the Southeast specializes on them—other snakes! Because snakes are long and thin and must swallow their prey whole, other snakes make ideal prey for many species. Typical snake predators in the Southeast include coral snakes, common kingsnakes, and indigo snakes. Some snake-eating snakes sometimes even swallow snakes larger than themselves. Snake-eating snakes can effectively target other snakes because they can pursue them in their favorite hiding places: underground burrows, beneath logs, and in rock crevices.

Snake eggs are vulnerable to a wide variety of animals. Scarlet snakes specialize on reptile eggs, and some herpetologists consider nonnative fire ants to be a major threat to the eggs of species such as the southern hognose snake.

Although they are venomous, copperheads still use camouflage to hide from predators.

Death feigning is a defensive response of hognose snakes.

DEFENSE

Southeastern snakes exhibit remarkable—and sometimes very entertaining—responses to threats from predators and people. The first level of defense for most southeastern snakes is to go unseen, which they achieve by hiding out of sight or being effectively camouflaged. Once discovered, the initial response of most species, including venomous ones, is to flee to safety.

When a clean escape seems unlikely, a snake may try to fool or distract a predator. Many try to bluff their way out of the situation. The most common approach, used by many species, is to make the head and body look bigger. Some harmless watersnakes flatten and expand their heads and even their entire bodies so that they resemble a broad-headed, thick-bodied cottonmouth. Thus, venomous snakes cannot be distinguished from harmless species simply because they have large, "arrow-shaped" heads.

When confronted or captured, many southeastern snakes release foul-

The classic open-mouthed defensive display of the cottonmouth is often sufficient warning to deter predators.

Many harmless snakes such as the eastern hognose and most watersnakes expand their head and neck regions to appear bigger and more threatening.

The characteristic defensive rattle cannot be heard by the snake itself but serves as a warning to would-be predators.

The defensive display of a ringneck snake

smelling musk from the cloaca. Each species of snake has its own odor, ranging from the sickeningly sweet, almost perfumed scent of garter snakes to the nauseatingly thick and overpowering musk of large watersnakes. Some species, such as eastern diamondback rattlesnakes, will even spray their musk a short distance if captured. The smells released by snakes presumably discourage some predators, which would prefer not to eat something that smells (and presumably tastes) so bad.

Other defensive displays include the red-bellied snake's curling of the upper lip to reveal what appear to be large teeth; the pine snake's open-mouthed hissing; the cottonmouth's gaping; and the death feigning of several species, with hognose snakes giving the most spectacular (and realistic) performance.

Many harmless snakes will bite when threatened.

Mud snakes, ringneck snakes, and coral snakes unable to escape a threat sometimes put the head under the body, curl the tail, and turn it upside down, exposing the brightly colored undersurface. Whether the display is a threat or an attempt to divert the attack away from the head and toward the tail is unknown.

The last resort for many snakes is to bite an attacker, although many will not bite a person under any circumstances. A bite can have serious consequences for people or predators if the snake is venomous. Bites from the larger watersnakes, racers, and rat snakes can draw blood, but typically the injury is equivalent to a scratch and requires no more than simple cleaning with soap and water, and possibly the application of an antiseptic. A bite from a pit viper or coral snake can be painful and even dangerous depending on the location of the bite and the amount of venom injected. However, even for venomous snakes, biting is a last resort.

Some defensive behaviors are used in response to specific predators. For example, kingsnakes are immune to the venom of pit vipers. Rather than trying to defend itself by biting, a pit viper attacked by a kingsnake will arch its back toward its attacker.

REPRODUCTION

Snakes are similar to mammals and birds in their general reproductive patterns. Males mate with females by means of internal fertilization, which is sometimes preceded by elaborate courtship behaviors that may include combat between adult males. Male-male combat among some of the large species of snakes (e.g., rat snakes, kingsnakes, cottonmouths, and rattlesnakes) is a rarely seen but presumably widespread behavior. During the combat, two males of the same species face each other, lift their heads above the ground, and entwine the front part of their bodies. The snake that can force its opponent to the ground wins the bout. Biting or injury is apparently rare in these combat encounters, even among venomous species. Because larger males generally win and get to mate with the female, males of species with male-male combat rituals are typically larger than females.

Baby red-bellied snakes (shown with their mother) are tiny.

Herpetologists have not observed the courtship of many species in the wild, but presumably it is a ritualistic activity that is distinctive for each species. Often the male crawls on top of or beside the female, ultimately placing his cloaca adjacent to hers. During courtship, snakes often twitch or jerk erratically, and in some species the male bites the female on the neck or head. The female generally signals her willingness to mate by lifting her tail and allowing copulation to occur. Garter snakes and some

Courtship behavior and actual mating by snakes are seldom seen in the wild. Kingsnakes generally mate as a single pair, whereas several male watersnakes may try to mate with one female.

watersnakes are noted for courtship behavior in which two or several males attempt to mate with a single female at once or in succession.

Most snakes use chemical signals called *pheromones* as part of their mating behavior. Generally, the female snake releases a pheromone as a signal to males that she is ready to mate.

Most species of southeastern snakes mate in the spring, although some mate primarily in autumn, and others may mate in both seasons. After mating, the female produces young either by laying eggs or by giving birth to live babies. Approximately half of southeastern snakes lay eggs, and half give birth to live young. Recent research has shown that some of the species that carry the young internally until birth actually nourish the developing babies through placenta-like structures. Other live-bearing species carry the developing young inside thin membranes similar to developing eggs. Most egg-laying snakes lay eggs in late spring or early summer, and the eggs hatch in mid-to-late summer or fall. Live-bearing snakes generally give birth during the same period.

The egg-laying snakes deposit groups, or *clutches,* of elongate, white or cream-colored eggs with leathery shells. The nesting sites are usually underground or beneath logs or rocks. Pine snakes tunnel several feet beneath the surface in sandy areas to create a cavity for their eggs. Rat

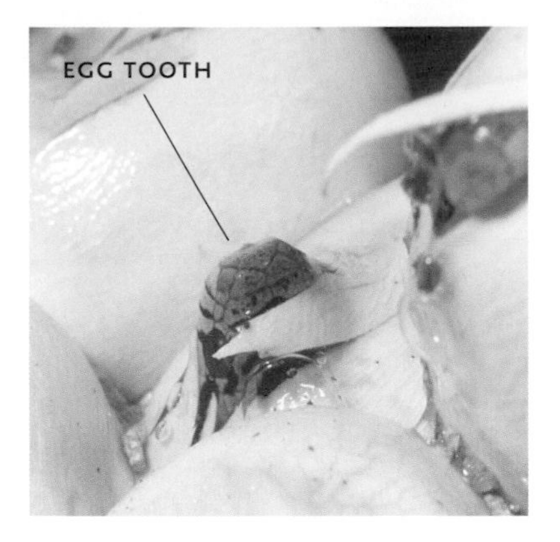

Baby snakes use an egg tooth to slice open the egg from the inside.

snakes will often lay eggs under rotting vegetation in the hollow of a tree. The incubation period ranges from an average of only a few days or weeks for some species to more than 2 or 3 months for others. The temperature and humidity affect the success and rate of hatching. Baby snakes have an egg tooth on the tip of the snout that allows them to slice through the leathery eggshell. The egg tooth is lost the first time the young snake sheds its skin.

The females of some snakes, such as mud snakes and rainbow snakes, are believed to remain with the eggs until they hatch. Some pit vipers (live-bearers) protect their newborn young for several hours or days. Some species may exhibit complex parental behavior, but it has rarely been seen.

Some snakes, including rattlesnakes, give live birth; others, such as worm snakes, hatch from eggs.

Rough green snakes are often found in shrubs and vines.

LOCOMOTION

How do snakes move about without limbs? In fact, they get around very well. The coordinated operation of muscles, flexible ribs, and overlapping belly plates (the scales on the snake's belly) allows snakes to maneuver down tunnels, over the ground, up trees, and in the water. The coordination of ribs and belly plates is especially important in allowing a snake to push itself along the ground or a tree limb. Each belly plate is associated with the ends of a pair of ribs. The most common method southeastern snakes use to move forward is called *lateral undulation*. As the long muscles down the body of the snake are contracted first on one side and then on the other, the ribs and belly plates push backward and the snake moves forward. The alternation of contractions on the two sides of the body is so well synchronized that snakes appear to glide effortlessly over the ground, across bushes, or into holes. Swimming snakes undulate their entire bodies to push themselves forward in the water.

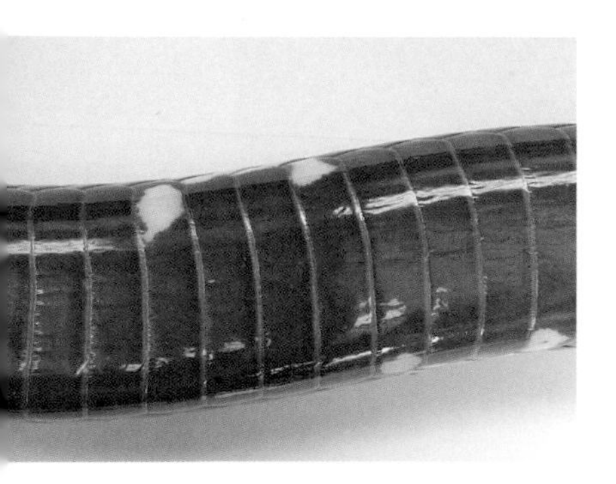

Belly plates (ventral scales) coordinate with the ribs and help to push a snake forward.

Southeastern snakes occasionally use other forms of locomotion more common in snakes from other regions. Racers will use a *sidewinding* motion to cross a hot highway, for example, the same way sidewinder rattlesnakes of the Southwest move across sandy deserts. A sidewinding snake lets only the front and back ends of its body touch the ground, pushes itself up vertically, and then moves its body horizontally. Large rattlesnakes often use *rectilinear locomotion* when crawling overland or across highways. In this form of movement, muscles on both sides of the body contract simultaneously so that the snake slowly moves forward in a straight line. Snakes using *concertina locomotion* extend the front part of the body forward, anchor it, and then pull the hind end forward. Most snakes use concertina locomotion when in burrows, and rat snakes typically use it when climbing.

Their lack of legs does not make snakes less efficient at locomotion than other animals. The amount of energy a snake uses to move a given distance is similar to that a lizard uses to go the same distance using its legs. Despite their efficiency in traveling over the ground, climbing, and swimming, snakes do not move as rapidly as they sometimes appear to. Black racers, for example, are among the fastest snakes in the Southeast, but their top speed when crossing a highway is typically less than 5 miles per hour. They probably can move faster in a natural habitat. But no snake can move over the ground faster than the average person can run.

Did you know?

All snakes can swim, even rattlesnakes. However, except for the aquatic species of snakes most stay on the surface of the water and do not dive.

Rat snakes can climb straight up the side of a tree or even up a brick wall.

An x-ray of a cottonmouth with a surgically implanted radiotransmitter and PIT tag

ACTIVITY

Seasonal Activity

Our impression of when snakes are active is influenced by those we see most often. The species that travel overland to get from one wetland habitat to another, to reach hibernation sites, or to search for prey or for mates are the ones people are most likely to see. Snakes and other reptiles become inactive during extremely cold and hot periods because they cannot maintain their body temperature at a suitable level. Thus, they generally become dormant during cold winters (*hibernation*; some scientists refer to this as *brumation*) and hot, dry summers (*aestivation*). Because most of the Southeast experiences moderate temperatures in most years and most seasons, some snakes are active to some degree during every month in most areas. Southeastern snakes are dormant for longer periods during the winter at higher latitudes and elevations because seasons are shorter and colder there, but they are more likely to be active throughout most of the summer in such areas. Watersnakes and cottonmouths may aestivate during the hottest and driest parts of the summer when their prey is scarce.

Springtime is generally the period of greatest activity for southeastern snakes as a group, both because many species breed in the spring and also because they are seeking meals as they emerge from winter dormancy. Because eggs hatch and live-bearing snakes give birth during late summer and fall, however, more snakes are present and likely to be seen aboveground then. The young of some species such as hognose snakes, racers, and garter snakes are especially common during autumn. Most of these young snakes are eaten by predators and do not survive until the following spring.

The timing of courtship and breeding influences the aboveground activity of snakes during some seasons. Thus, pine snakes are encountered most frequently in the spring, presumably because that is their breeding season and they are actively pursuing mates. In the spring and fall, snakes are often seen crossing highways, although many do not survive the trip. Canebrake rattlesnakes, especially large males, are commonly killed crossing roads during their late summer and fall breeding season.

Some of the larger species of snakes are more often seen in the fall and spring as they enter and leave their winter hibernation sites. In the mountainous regions of the Southeast, some species gather in large numbers to hibernate in dens, which are usually in a rocky outcrop on a sunny south- or west-facing slope. Rattlesnakes, copperheads, and rat snakes frequently den together, and some individuals travel miles to reach a particular denning site. Garter snakes are noted for hibernating together in large numbers in some regions. The same snakes generally return to the same dens year after year.

Hognose snakes search for toads during the daytime.

Regional temperature patterns differ from year to year, and snake species respond to these annual variations in different ways. Some that disappear underground and become dormant during extended periods of winter cold will remain active during more moderate winters, occasionally appearing aboveground. For example, cottonmouths will sometimes emerge from underground hibernating sites to warm up in the winter sun, even on very

Scarlet snakes are active only at night.

Watersnakes are often found out during the daytime basking.

cold days. Canebrake rattlesnakes living in the same region do not typically emerge from hibernation until mid-to-late spring. Likewise, some snakes become active earlier in the year during a warm spring, while others do not emerge from their winter dormancy until late spring regardless of the spring temperatures in a particular year.

Daily Activity

Snakes are active during both day and night in the Southeast, but daily activity periods vary from one species to the next. Pine snakes and racers, for example, are *diurnal*; that is, they are characteristically active aboveground only during the daylight hours. Scarlet snakes, on the other hand, are *nocturnal*; that is, they are found out only at night. Some species (e.g., corn snakes and copperheads) are typically active during the day in the cool seasons and at night during summer. Many watersnakes are active around water at night when searching for nocturnal prey such as frogs, but may also be active during the day.

Underground is one of the safest places for a snake, and many species spend most of their lives in root tunnels or beneath logs, rocks, and ground litter. All southeastern snakes are adept at using underground pathways, and some, such as pine snakes and hognose snakes, make their own burrows in sandy or other loose soils. A behavior that tends to make snakes highly visible is basking in the sun, which snakes do to warm themselves. Watersnakes are especially noted for basking on limbs or bushes above water, into which they quickly retreat if disturbed. Likewise, they will often rest on shaded limbs during warm days. Many terrestrial snakes bask on the ground or on rocks during cool periods, often close to a hole or other retreat.

Did you know?

Our perceptions of activity are limited for some snake species because many are more active beneath the surface than aboveground.

Garter snakes are often active during cool weather.

TEMPERATURE BIOLOGY

Temperature affects nearly every aspect of the lives of snakes. It can influence a snake's growth, capture of prey, and ability to escape from predators. A snake's body temperature is determined primarily by the environmental conditions that surround it, and thus most people refer to them as *cold-blooded*, although their blood is not necessarily cold. Scientists refer to cold-blooded animals as *ectotherms*, but both terms mean the same thing.

Coachwhips prefer higher temperatures than some snakes and are usually active on warm days.

Because of their low-energy lifestyle, many snakes, such as this eastern diamondback rattlesnake, need to eat only a few meals each year.

Glossy crayfish snakes sometimes bask on sunny days during late winter.

Most snakes prefer to maintain a body temperature around 86° F (30° C). Humans, other mammals, and birds regulate their body temperature by using internal body heat, which they produce during metabolism. Snakes, however, regulate their body temperature by using different behaviors—such as basking on cool, sunny days and seeking shelter and shade on hot days. They are particularly susceptible to heat stress.

Low-Energy Lifestyle

Just as there are certain advantages to being warm-blooded, there are also advantages to being cold-blooded, or ectothermic. Ectotherms have low metabolic rates and thus generally require only about one-tenth the food needed by a similar-sized mammal. This low-energy lifestyle allows snakes to go for long periods between meals and to specialize on food that is available at only certain times of the year. A good example is the egg-eating scarlet snake, which feeds primarily on lizard eggs during late spring and summer. Because snakes do not use metabolic heat to maintain a high body temperature, like mammals and birds do, they can use much more of the energy from their food for growth and reproduction.

IDENTIFYING SNAKE SPECIES OF THE SOUTHEAST

Being able to identify animals and plants is one of the first steps in developing an appreciation for nature. When you know what species a snake is, you can find out what it might eat, how large it can get, and whether it is venomous or harmless. Most of the more than 50 species of snakes native to the Southeast can be identified using a few key characters, especially when coupled with the geographic location (e.g., a snake native to Alabama or Florida with a bright green body is without question a rough green snake, and a shiny black snake with red stripes down its back and sides from coastal Virginia or southern Mississippi will always be a rainbow snake). Such information is easy to learn and can greatly add to your enjoyment of nature in general and snakes in particular.

Body scale counts and head scale configurations are among the key characters used to tell snake species and subspecies apart, but these are subtle features used primarily by professional herpetologists. Other traits are obvious to even the most casual observer. For example, only three southeastern snake species have rattles at the end of the tail. The following guide to the species presents a combination of characters, some quite obvious and others requiring closer examination, that will distinguish each species from all or most of the others. A very few species will never be easily told apart without careful examination by an expert in herpetology. And even professional herpetologists will sometimes have difficulty confirming whether the small brown snake in their hand is a smooth or rough earth snake unless they are holding a magnifying glass. But most southeastern snakes can be correctly identified with minimal effort by using this book.

Did you know?

Although albino snakes are commonly bred in captivity, they are extremely rare in the wild.

Colors and Patterns

Southeastern snakes exhibit an amazing diversity of patterns, ranging from solids to blotches, spots, and bands or rings to longitudinal stripes. The most common colors are blacks, browns, and grays, which offer protection through camouflage. More than 20 of our snake species, however, are brightly colored in shades of red, orange, yellow, or green. The color and pattern of some species—such as rainbow snakes, green snakes, and southern hognose snakes—rarely vary and can be used with confidence in identification throughout the Southeast (see the individual species accounts for descriptions). Other species, such as garter snakes, red-bellied snakes, and eastern hognose snakes, exhibit variation—even at the same locality—that makes color and pattern unreliable as identifying characters. In some species, intergrades that possess color patterns and scale counts intermediate between two or more subspecies may occur where the geographic ranges of the subspecies overlap. All species exhibit aberrant color patterns on rare occasions (e.g., solid black copperheads and *albino* rattlesnakes).

Size as a Character in Identification

Southeastern snakes range in body size from the tiny worm snakes and earth snakes to the enormous pine snakes and indigo snakes, which may weigh a thousand times more than the smaller species. Because it is a straightforward and easy measurement, body length is the standard method of determining a snake's size, although body shape and bulk can also be important factors for identification. A 2-foot-long southern hognose snake, for example, will weigh much more than a ribbon snake of the same length. It is important to remember than every snake will change greatly in length during its lifetime (most will be five times longer as adults than their size at birth), and many species vary in body proportion based on their sex and recent feeding success. Therefore, body size is not always a reliable character. It may be worthless for differentiating between some species and highly reliable for separating others. For example, a baby coachwhip

Did you know?

Four native species of southeastern snakes (eastern and western worm snakes, mud snake, and rainbow snake) have a pointed spine on the tip of their tails.

In the Southeast, only rattlesnakes, copperheads, and cottonmouths have a heat-sensitive pit between the eye and nostril.

> **Note** Some of the characters used to distinguish venomous snakes may be difficult to see from a safe distance. The best way to identify whether a snake belongs to a venomous species is to learn what each of the venomous species in your region looks like.

or pine snake is longer at birth than a crowned snake or pine woods snake will ever be.

Another size trait sometimes useful for distinguishing among species is the length of the tail in relationship to the total body length. For example, rattlesnakes have very short tails relative to their body length (less than 10%), while the tail of ribbon snakes can be more than a third of the total body length. In all southeastern snakes, the male's tail, from birth through adulthood, is always proportionately longer than a female's of the same species.

In the southeastern United States, venomous pit vipers characteristically have a single row of scales beneath the tail (top) whereas all harmless species (as well as the venomous coral snake) have a double row of scales (middle and bottom respectively). The trait is recognizable on shed skins and can sometimes be useful in determining whether a snake is venomous.

How Can You Tell If a Snake Is Harmless or Venomous?

People ask more questions about the identification, behavior, and ecology of snakes than about any other group of animals in the Southeast. One of the first questions a person wants to know on finding a live or dead snake, or even a shed skin, is whether the snake is venomous. All six of the venomous species found in the Southeast differ from all of the harmless species by having fangs in the front part of the mouth, although it is not usually necessary (or recommended) to open a snake's mouth to determine its identity. All five of the southeastern pit viper species (rattlesnakes, cottonmouth, and copperhead) have three distinctive characteristics that no other snake in the region has: (1) there is an opening (the heat-sensitive pit) in the side of the head between the nostril and the eye; (2) most of the scales on the underside of the tail are in a single row rather than in a double row (see photos); and (3) the pupils in the eyes of pit vipers appear elliptical in most situations, whereas all other southeastern snakes (including coral snakes) have distinctly round pupils.

A simple rule learned as a rhyme by children in the Southeast readily separates an eastern coral snake from all of the harmless snakes of the region: Red touch yellow, kill a fellow—red touch black, venom lack. That is, any southeastern snake with brightly colored red, yellow, and black rings in which the wider red rings touch the narrower yellow ones is almost certainly a coral snake. The outlines of the rings are visible even on a shed skin.

One common bit of erroneous information that can cause a nonvenomous snake to be misidentified as venomous is that only rattlesnakes vibrate their tails. Many harmless snakes, especially large terrestrial species, vibrate the tail when confronted, and in dry leaves may actually sound like a rattlesnake. Also, a large, triangular head does not necessarily identify a snake as venomous. Many of the large but harmless watersnakes flatten and expand the head when alarmed, making them look very similar to the venomous cottonmouth.

Key Traits to Look for in Identifying Snakes

A few characteristics can be used to narrow the field by excluding certain species.

SCALE TYPE SMOOTH OR KEELED (SEE PHOTOS) One of the most reliable traits for most southeastern species is whether the body scales are mostly *keeled* (i.e., with a ridge down the center that makes them appear rough) or mostly *smooth*. For example, all of the pit vipers and most of the watersnakes have keeled scales, and more than 20 of the harmless snake species and the coral snake have mostly smooth scales. A few species, such as rat snakes and corn snakes, have weakly keeled scales, and the males of some smooth-scaled species, such as indigo snakes and striped crayfish snakes, have a few keeled scales at the base of the tail; indigo snakes may even have weakly keeled scales on the back. The imprint of the keel, when present, is visible on each scale of a shed skin, too.

Most snake species have either smooth or keeled (rough) scales.

ANAL PLATE In most snake species, the *anal plate*—the last belly scale, which covers the cloaca and precedes the tail scales—is either whole and undivided or divided into two scales. The trait can be especially useful when you are identifying a shed skin.

BODY SHAPE Snakes vary in shape from species that are characteristically robust for their length to those that are usually very slender. Like other biological traits, shape can vary considerably within a species (e.g., although cottonmouths are typically heavy bodied, thin individuals are occasionally encountered), but most species are relatively consistent in their overall body shape.

PATTERN AND COLOR The presence or absence of encircling rings, bands, blotches, or longitudinal stripes can be used to distinguish many species, although in some species appearance can vary significantly from one individual to the next. For example, garter snakes usually have three yellow stripes, but in some areas a checkered pattern predominates and the stripes are indistinct. Also, some watersnakes, cottonmouths, and eastern hognose snakes turn darker with age, so that the body markings of larger individuals may not be visible. Likewise, the young of some of the larger species look very different from the adults. A certain amount of experience is necessary to recognize snakes at first sight.

DISTINCTIVE CHARACTERS Special traits or characteristics can often be useful in narrowing the field to only a few possibilities. Behavior can be an instant clue to a snake's identity; for example, the open-mouthed gape of a cottonmouth, the rattling of a rattlesnake (not to be confused with the tail vibration of harmless species), or the cobralike display of a hognose snake. A large southeastern snake with a hardened, spiny tail tip will be a mud snake or rainbow snake. A distinctive color pattern of red, yellow, and black rings should serve immediate notice that the snake is a coral snake or one of the harmless mimics (scarlet kingsnake, scarlet snake, or Louisiana milksnake, see page 213). Knowing the distinctive behaviors, appearances, and structural characteristics of the different species can be a useful tool in identifying snakes in the wild.

GEOGRAPHIC LOCATION Where a snake is found in the wild usually provides an instant clue about its identity. Copperheads and canebrake rattlesnakes are not found in the southern half of Florida; diamondback and pigmy rattlesnakes are absent from Virginia. Identification of a local species can often be narrowed considerably by eliminating species that do not naturally occur in the region.

HABITAT The habitat where the animal was found can be another important clue to its identity. Although exceptions always exist, most of the watersnakes are unlikely to be found far from water, and typical sandhills species such as southern hognose snakes and pine snakes are unlikely to be found near water. Even the specific location within the habitat may provide meaningful information. For example, a coral snake or ringneck snake would not be expected to be in a bush or tree.

TIME OF DAY The active period of a species is often restricted to daytime or nighttime. A snake found crawling around in the day is highly unlikely to be a scarlet snake, and a snake found crossing a road at midnight is not going to be one of the hognose snakes.

Did you know?

Florida has 43 species of native snakes, more than any other southeastern state. Alabama, Georgia, Louisiana, and Mississippi each have 41. Virginia has the fewest snake species (30).

Salt marsh snakes in Florida are often associated with mangroves.

Species Accounts

INTRODUCTION

The following species accounts are designed to help the reader become familiar with every species of snake native to the Southeast. The accounts are grouped into five categories based on body size and ecology rather than arranged in the traditional groupings seen in most field guides. The five categories are (1) small, (2) mid-sized, and (3) large terrestrial snakes; (4) snakes associated with aquatic habitats; and (5) snakes that use venom to kill their prey and are potentially dangerous to humans. While these groupings are somewhat subjective, and the size ranges of juvenile mid-sized snakes often overlap those of small snakes, they are easily understandable and provide an approach that will help people to become aware of the similarities and differences among the snakes of the Southeast.

Each species' geographic range in the Southeast is indicated on the map that accompanies each account. A smaller map shows the entire U.S. range for species that occur outside the Southeast. The range maps are based on historical records, although ranges have changed for many species in recent years as snake populations have declined. The shaded range should be viewed as an approximation of the actual presence of a species, as almost no snake has a continuous distribution across all habitats within a region. As an example, the range map for ringneck snakes indicates that they occur throughout the Southeast. Their actual distribution is patchy, however, as a result of their habitat requirements and the disappearance of populations due to natural or human-based causes.

HOW TO USE THE SPECIES ACCOUNTS

The species entries are arranged as follows (not all elements occur with every entry):

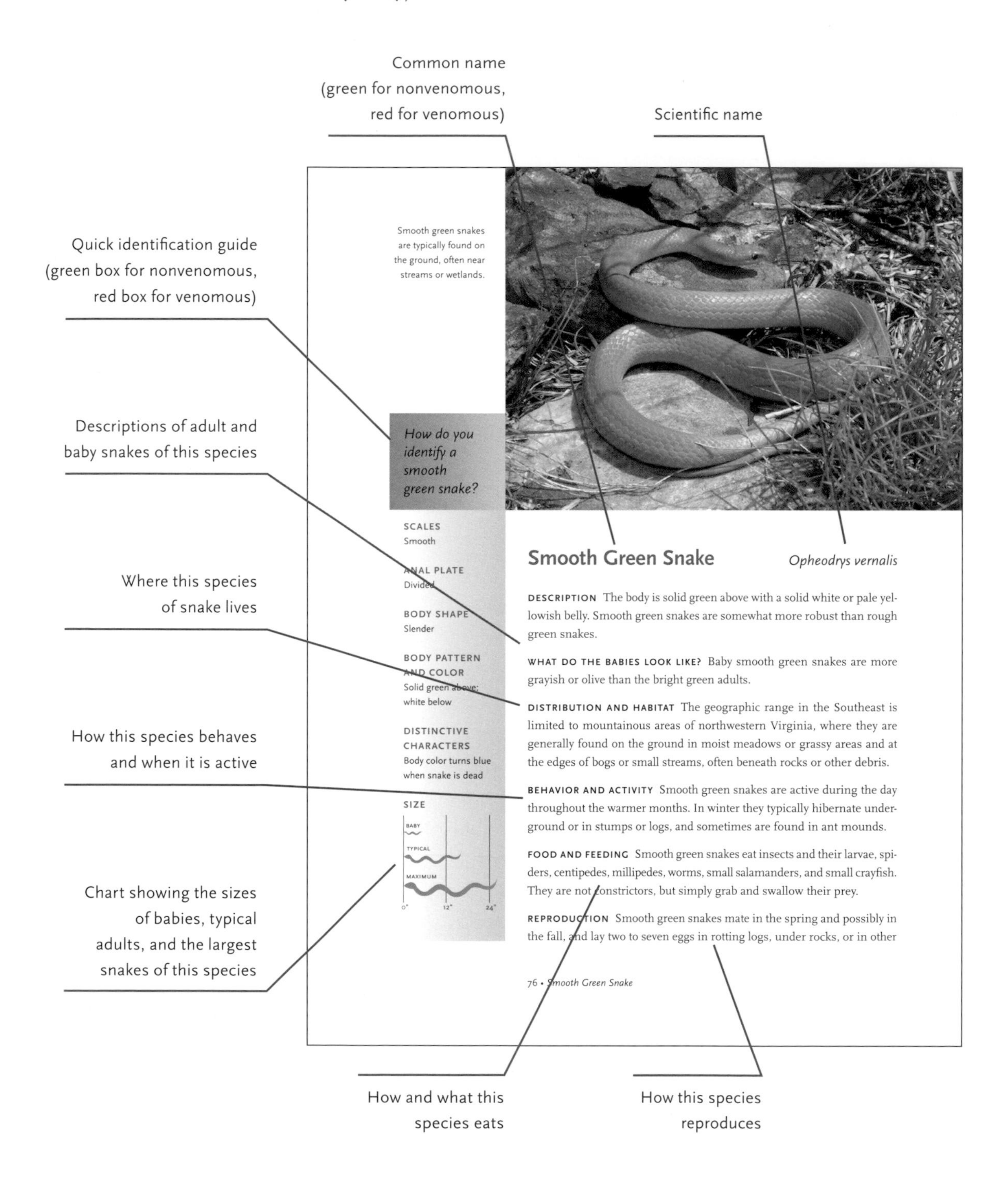

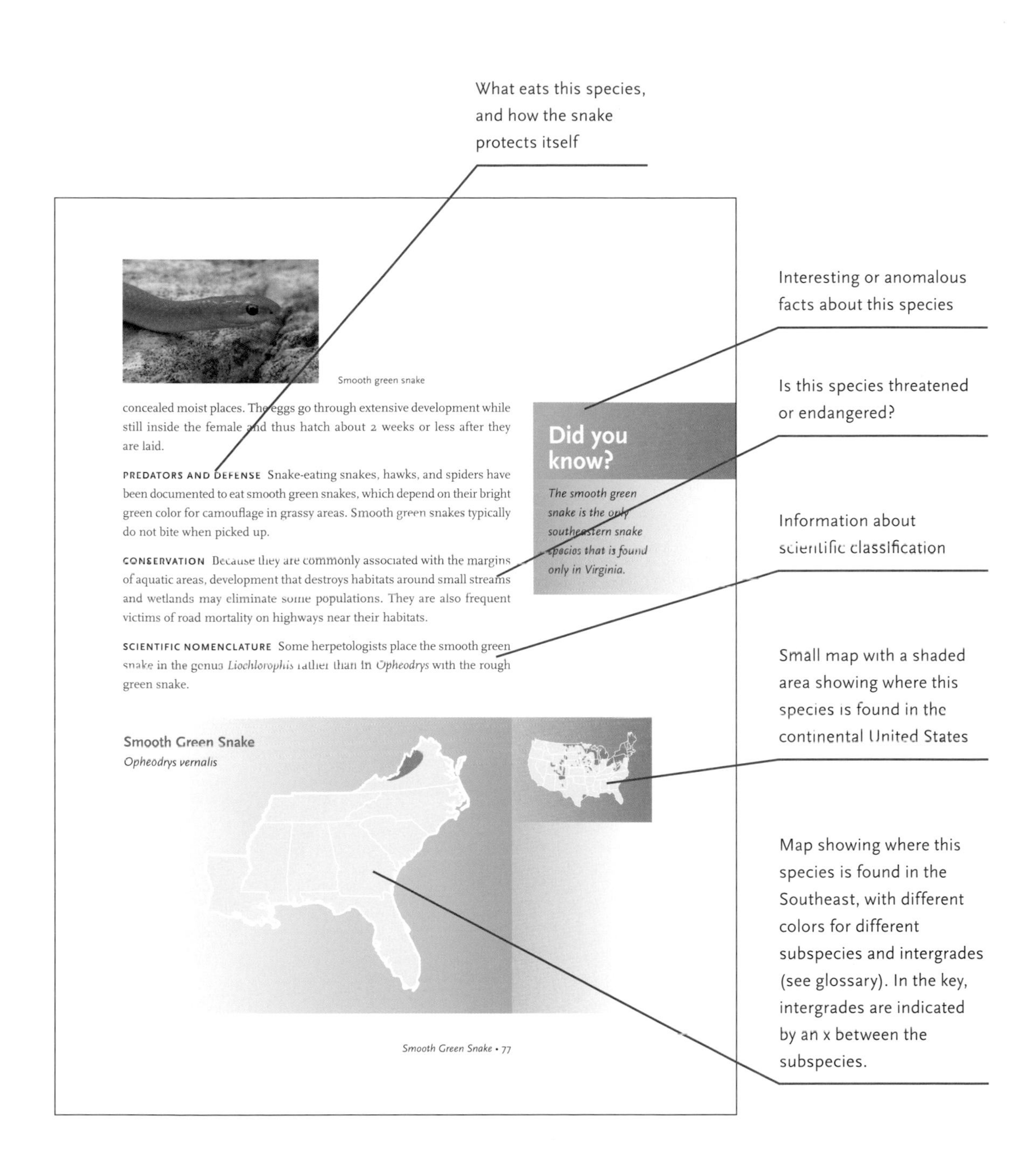

What eats this species, and how the snake protects itself
Interesting or anomalous facts about this species
Is this species threatened or endangered?
Information about scientific classification
Small map with a shaded area showing where this species is found in the continental United States
Map showing where this species is found in the Southeast, with different colors for different subspecies and intergrades (see glossary). In the key, intergrades are indicated by an x between the subspecies.
Smooth green snake
concealed moist places. The eggs go through extensive development while still inside the female and thus hatch about 2 weeks or less after they are laid.
PREDATORS AND DEFENSE Snake-eating snakes, hawks, and spiders have been documented to eat smooth green snakes, which depend on their bright green color for camouflage in grassy areas. Smooth green snakes typically do not bite when picked up.
CONSERVATION Because they are commonly associated with the margins of aquatic areas, development that destroys habitats around small streams and wetlands may eliminate some populations. They are also frequent victims of road mortality on highways near their habitats.
SCIENTIFIC NOMENCLATURE Some herpetologists place the smooth green snake in the genus Liochlorophis rather than in Opheodrys with the rough green snake.
Did you know?
The smooth green snake is the only southeastern snake species that is found only in Virginia.
Smooth Green Snake
Opheodrys vernalis
Smooth Green Snake • 77

SMALL TERRESTRIAL SNAKES

Rough earth snake

How do you identify an earth snake?

SCALES
Smooth earth snakes, smooth scales; rough earth snakes, keeled scales

ANAL PLATE
Usually divided

BODY SHAPE
Moderately stout to stout

BODY PATTERN AND COLOR
Solid brown above with whitish belly

SIZE

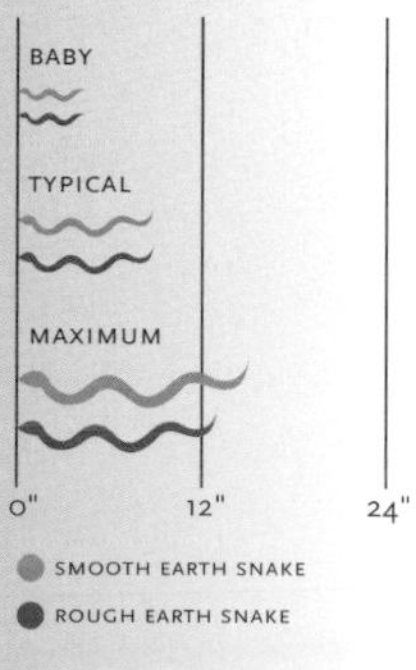

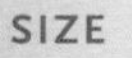

Smooth Earth Snake *Virginia valeriae*
Rough Earth Snake *Virginia striatula*

DESCRIPTION The earth snakes are small, brown to brownish gray snakes with light-colored bellies. Their heads are small and their noses are somewhat pointed. The rough earth snake (*V. striatula*) has keeled scales and sometimes has an indistinct light band across the back of the head. The smooth earth snake (*V. valeriae*) has smooth scales and often has a scattering of tiny dark spots on the back. Sometimes these dots are arranged into faint stripes. Three subspecies of the smooth earth snake are recognized (eastern smooth earth snake, *V. v. valeriae*; western smooth earth snake, *V. v. elegans*; and mountain smooth earth snake, *V. v. pulchra*), but the differences among them are rather obscure.

WHAT DO THE BABIES LOOK LIKE? Baby earth snakes are similar to the adults.

Smooth earth snake

DISTRIBUTION AND HABITAT Earth snakes are found in parts of all the southeastern states in pine or hardwood forests, including the margins of swamps and other wetlands, and in fields and rural areas adjacent to woodlands. They are sometimes common in suburban areas.

BEHAVIOR AND ACTIVITY Unless temperatures are extremely cold, earth snakes can be active year-round, including during warm spells in winter, but they usually remain beneath ground litter rather than on the surface. They are active aboveground primarily in early evening or at night during hot times of the year.

FOOD AND FEEDING Both smooth and rough earth snakes eat mostly earthworms. They may also eat the adults, larvae, and eggs of small, soft-bodied insects, including ants and termites, as well as slugs and small snails. Earth snakes simply grab their prey and swallow it while it is still alive.

Some rough earth snakes have a light band across the back of the head.

REPRODUCTION Earth snakes mate in the spring and early summer, but nothing is known about their courtship. Presumably, mates find one another by following chemical trails. Earth snakes give birth to 2–14 (usually about 8) young during mid-to-late summer or early fall. Rough earth snakes nourish their young as they are developing inside the mother's body using a structure similar to the placenta of mammals.

Smooth earth snake

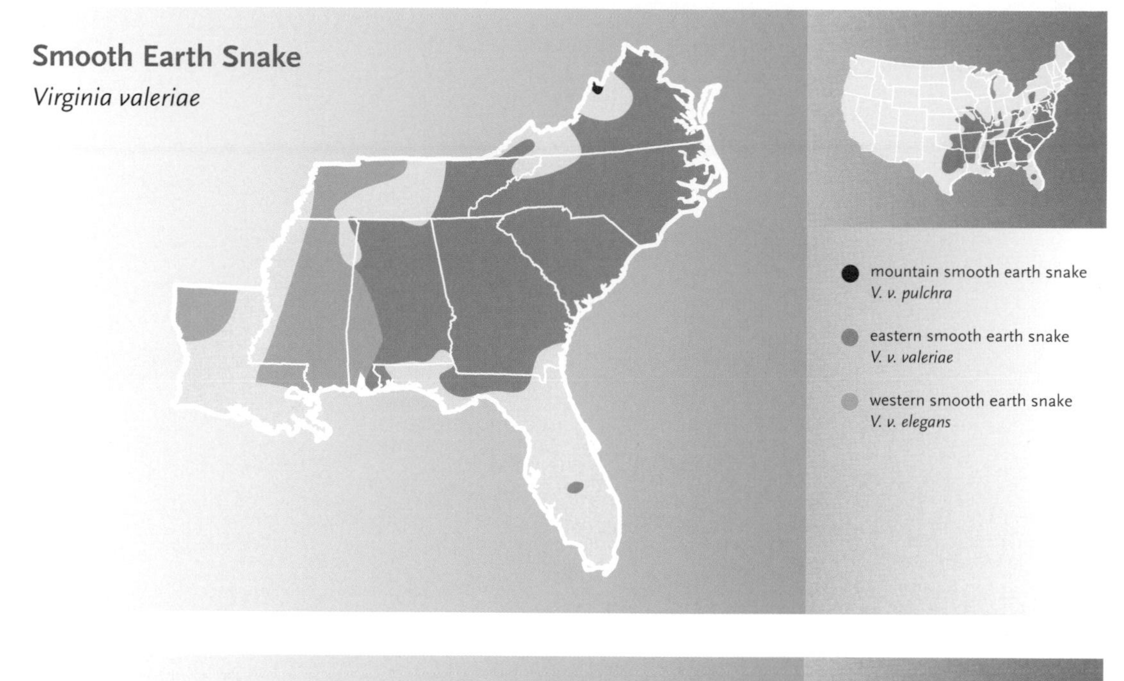
Smooth Earth Snake
Virginia valeriae
mountain smooth earth snake
V. v. pulchra
eastern smooth earth snake
V. v. valeriae
western smooth earth snake
V. v. elegans

Rough Earth Snake
Virginia striatula

Some smooth earth snakes have tiny dark spots on the back.

PREDATORS AND DEFENSE Because of their small size, earth snakes probably are eaten by a variety of animals; kingsnakes, coral snakes, and racers are known predators. Earth snakes may bite a predator when being attacked. They seldom if ever bite a human when picked up, although they may release musk from scent glands in the cloaca (anal region).

CONSERVATION No specific conservation threats are associated with earth snakes, although they are often common in suburban areas and many are killed by domestic cats.

Midland brown snake from Georgia

How do you identify a brown snake?

SCALES
Keeled

ANAL PLATE
Divided

BODY SHAPE
Moderately stout

BODY PATTERN AND COLOR
Two rows of dark spots on gray or brown body

SIZE

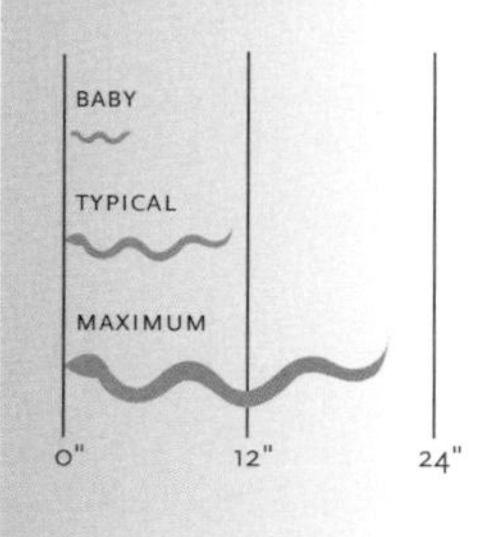

Brown Snake

Storeria dekayi

DESCRIPTION Brown snakes are grayish brown or brown above with two rows of small black spots running the length of the back. They usually have small, obscure dark spots on the sides as well. The belly is lighter in color with darker spots along the edges. Most individuals have a pair of larger spots on the neck that sometimes form a complete collar. Five subspecies occur in the Southeast (northern brown snake, *S. d. dekayi*; marsh brown snake, *S. d. limnetes*; Texas brown snake, *S. d. texana*; midland brown snake, *S. d. wrightorum*; and Florida brown snake, *S. d. victa*). The midland subspecies intergrades with the northern subspecies over a broad geographic area and with the Texas subspecies over a narrow band (see map). Only the Florida brown snake has a broad, light collar on the back of the head.

Baby brown snakes (shown with mother) have light neck rings.

WHAT DO THE BABIES LOOK LIKE? Baby brown snakes are similar to the adults but the body color is usually darker (often gray) with a pale yellow, grayish, or cream-colored neck ring.

OTHER NAMES This species is still called by its original name, DeKay's snake, by many herpetologists. That name sometimes caused problems in discussions with nonherpetologists who mistakenly heard the name as "decayed" snake.

DISTRIBUTION AND HABITAT This species is found throughout most of the Southeast in many types of woodland habitats, including hardwood and pine forests, elevated or seasonally dry areas in swamps, and the margins of wetlands. They are commonly found in wet floodplains and along creek bottoms. Brown snakes adapt well to many suburban areas and can be abundant in flower beds, grass clumps, and vacant lots. In some regions, brown snakes are actually encountered more frequently in residential areas and city parks than in natural forests.

Reddish phase of the brown snake from North Carolina

BEHAVIOR AND ACTIVITY Brown snakes hibernate during extreme cold spells in the winter but may be active anytime from early spring to late fall, and even during winter warm spells. They sometimes crawl aboveground as well as beneath leaf litter, pine straw, and grass or other surface vegetation and occasionally may be active during the day but are especially active at night.

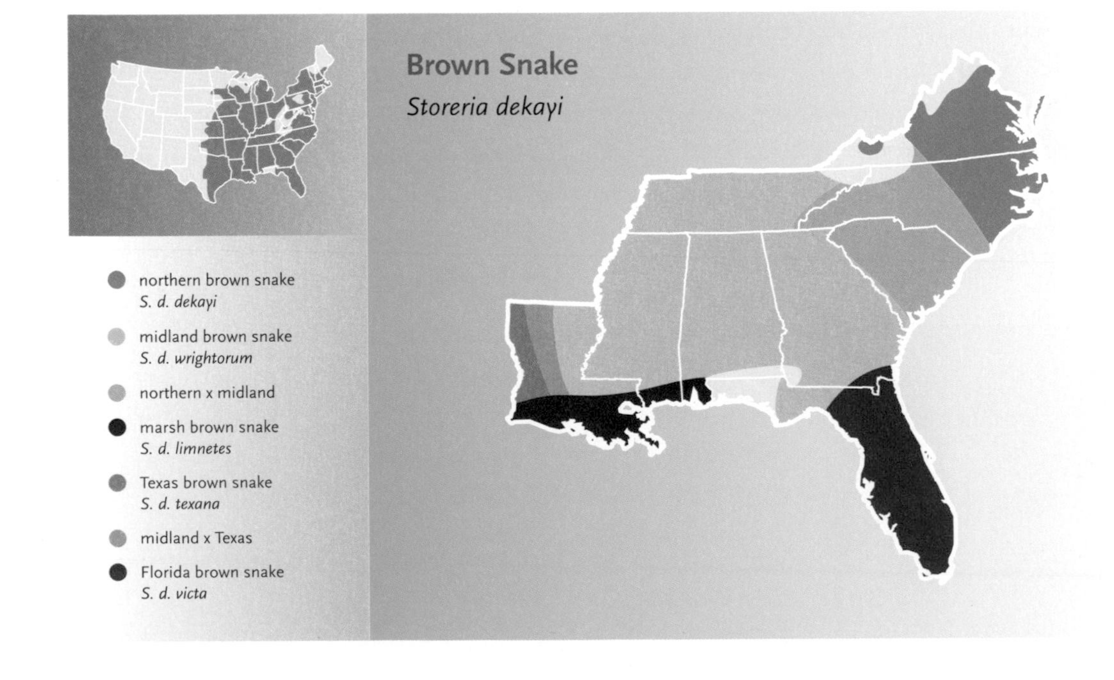

Marsh brown snake from southern Louisiana

FOOD AND FEEDING Brown snakes eat mostly earthworms and slugs, but will also eat snails, various insects, small salamanders, and spiders. They simply grab and swallow their prey alive. When feeding on snails, they pull the snail from its shell before swallowing it.

Earthworms are common prey of brown snakes.

REPRODUCTION Brown snakes mate in the spring (March–May) and give birth to 3–41 young (average about 13) during the summer or early fall. Farther south, mating and birth may occur even earlier. In contrast to most reptile embryos that are enclosed within a membrane and receive nourishment from yolk as they develop inside the mother, embryonic brown snakes may receive nourishment from a placenta-like structure similar to that of mammals.

PREDATORS AND DEFENSE Documented natural predators of brown snakes include mammals (e.g., shrews, raccoons, opossums, and skunks), birds (e.g., robins, shrikes, brown thrashers, and hawks), snakes (e.g., racers, kingsnakes, and garter snakes), and even toads and spiders. Brown snakes

Florida brown snakes may vary in general coloration but usually have a light collar on the back of the head.

The Texas brown snake is found in western Louisiana.

respond to threats in a variety of ways. They first attempt to escape; if that fails, they may play dead, hide the head beneath the body coils, or flatten the body to appear larger. Brown snakes are typically mild mannered when handled and usually release a foul musk from their anal glands but do not bite.

CONSERVATION Many herpetologists consider the Florida brown snake to be threatened in some areas due to extensive development within the species' restricted geographic range.

SCIENTIFIC NOMENCLATURE Some herpetologists consider the Florida brown snake to be a separate species (*S. victa*) rather than a subspecies of *S. dekayi*.

A typical red belly of the red-bellied snake

A red-bellied snake from Florida

Red-bellied Snake *Storeria occipitomaculata*

DESCRIPTION Red-bellied snakes are extremely variable in appearance. They may be gray, reddish, or brown, with three light spots on the neck often forming a light ring that contrasts with a darker head. The belly is typically solid red, reddish brown, or orange, and rarely an individual will have a nearly solid black belly. A light stripe may or may not run the length of the back. Two subspecies occur in the Southeast (Florida red-bellied snake, *S. o. obscura*; and northern red-bellied snake, *S. o. occipitomaculata*). The differences between the two subspecies are not very apparent, and a broad zone of intergradation occurs in the Southeast (see map).

WHAT DO THE BABIES LOOK LIKE? Baby red-bellied snakes are similar to the adults.

DISTRIBUTION AND HABITAT Red-bellied snakes are found in portions of all southeastern states in hardwood and pine forests, the margins of swamps and other wetlands, and marshy or boggy habitats. They seem to be less common in developed areas than their close relatives, the brown snakes.

BEHAVIOR AND ACTIVITY Throughout most of the Southeast, red-bellied snakes are active during all warm months and during warm periods in winter. During extended cold periods they hibernate underground or be-

How do you identify a red-bellied snake?

SCALES
Keeled

ANAL PLATE
Divided

BODY SHAPE
Slender to moderately stout

BODY PATTERN AND COLOR
Solid (sometimes with a stripe down the back) brown, black, gray, or orange above; belly red or orange

SIZE

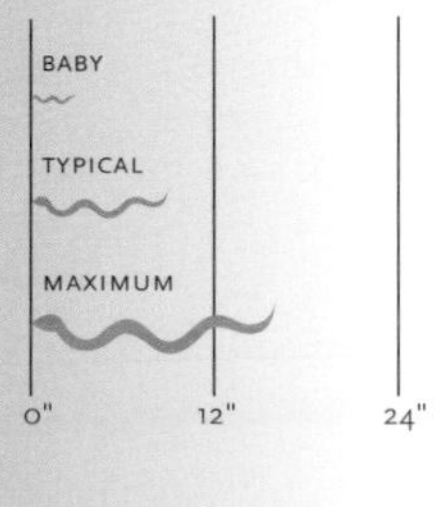

Most red-bellied snakes are light brown above, but some are dark and some have a light neck collar.

neath logs, rocks, and ground litter. Although they often remain hidden beneath leaves or other cover, they occasionally move about aboveground. Red-bellied snakes are active aboveground during early evening, or at night in the summer. They are noted for relatively long-distance travel. Some have been recorded moving several hundred feet—quite a distance for a small snake.

FOOD AND FEEDING Apparently, red-bellied snakes eat mostly slugs and earthworms, which they swallow alive. They may also feed on soft-bodied insects, snails, and isopods (pillbugs or "roly-polies").

REPRODUCTION Red-bellied snakes presumably mate in the spring but also have been observed mating during summer and fall. Nothing is known about their courtship. The female typically gives birth to 4–9 live young from May to June in Florida, and in late summer and fall in more northerly states. Rarely, litters may number up to 23.

PREDATORS AND DEFENSE Known or suspected natural predators of these small snakes are those characteristic of woodland habitats and include

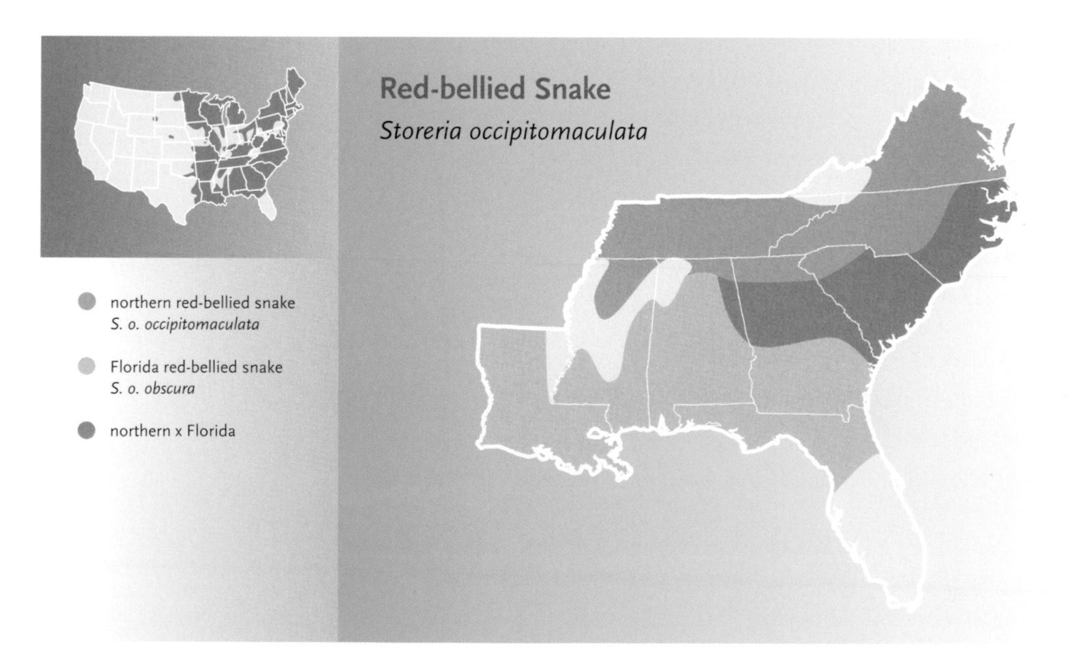

The Florida red-bellied snake (shown here) is not appreciably different in appearance from the northern subspecies.

snakes (e.g., kingsnakes and racers), birds, large spiders, toads, and salamanders. A peculiar response of many red-bellied snakes when they are picked up is to curl the upper lip to reveal a series of alternating black and white lines that look like large teeth and might appear threatening to a small predator. They may also respond with a combination of behaviors, including flattening the body, releasing musk, and playing dead. Red-bellied snakes are generally inoffensive toward humans and seldom if ever bite when handled.

CONSERVATION A major threat to red-bellied snakes in the Southeast and elsewhere is the destruction of woodland habitats. Domestic cats are also known to kill these small snakes in suburban communities.

How do you identify a southeastern crowned snake?

SCALES
Smooth

ANAL PLATE
Divided

BODY SHAPE
Slender

BODY PATTERN AND COLOR
Tan or grayish brown; head and neck black with a light band just behind the head

SIZE

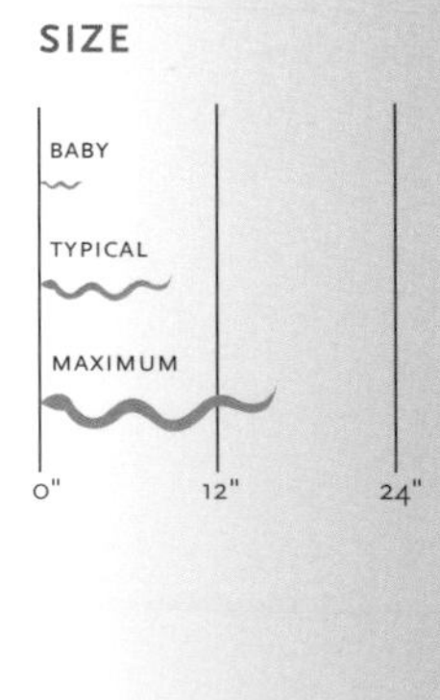

Southeastern Crowned Snake *Tantilla coronata*

DESCRIPTION This small, slender snake has a solid light brown or grayish brown back that is separated from the black head and neck by a light, cream-colored band. The belly is solid white or light pink. The head is rather pointed, and the lower jaw is countersunk.

WHAT DO THE BABIES LOOK LIKE? Baby southeastern crowned snakes look like the adults.

DISTRIBUTION AND HABITAT Southeastern crowned snakes are found in portions of all southeastern states but are replaced in peninsular Florida by other species of crowned snakes. They are generally most abundant in areas with sandy or other loose soil types and abundant surface litter, and may be found in both dry and moist woodland habitats.

BEHAVIOR AND ACTIVITY Southeastern crowned snakes hibernate during winter but are likely to be active beneath ground litter on warm days in early spring and late fall. Most aboveground travel is during early evening or at night, but individuals are active beneath logs, rocks, and ground litter during the day in spring, summer, and fall. Crowned snakes are adept burrowers in sandy soil, almost appearing to swim into the sand when discovered.

Southeastern Crowned Snake

Tantilla coronata

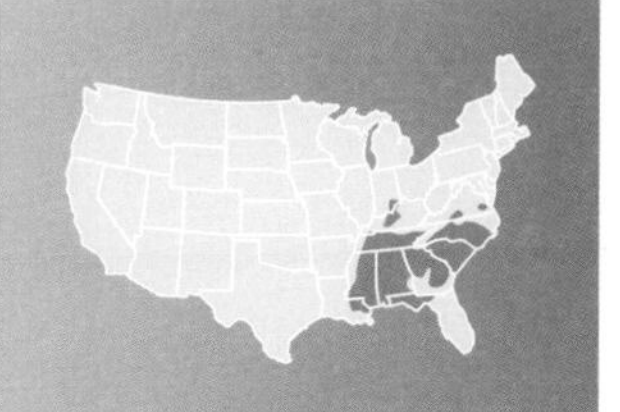

Centipedes are a common prey item of crowned snakes.

FOOD AND FEEDING Southeastern crowned snakes eat a variety of prey, including centipedes, insect larvae, earthworms, and spiders. Small, grooved fangs at the back of the jaw are used to direct venom into the prey. The venom's effectiveness in subduing prey is questionable, however, and all crowned snakes are presumably harmless to humans.

REPRODUCTION Mating occurs in spring, summer, and fall. Females that mate in the fall store the sperm until the following spring, when they produce one to three eggs in June and early July. The eggs hatch in the fall.

PREDATORS AND DEFENSE Because of their small size, crowned snakes are probably preyed on by most carnivorous vertebrates that occur in forested habitats, especially kingsnakes and coral snakes. Southeastern crowned snakes try to escape by burrowing into sand or soft soil, or by crawling beneath leaves or other ground litter. When captured, they do not bite but may release small quantities of musk from scent glands in the cloaca (anal region).

A light band separates the black head and neck of crowned snakes.

CONSERVATION Except for degradation and destruction of their forest habitats, southeastern crowned snakes face no particular conservation threats.

Rim rock crowned snake from southern Florida

How do you identify Florida and rim rock crowned snakes?

SCALES
Smooth

ANAL PLATE
Divided

BODY SHAPE
Slender

BODY PATTERN AND COLOR
Tan to reddish brown with a dark head

SIZE

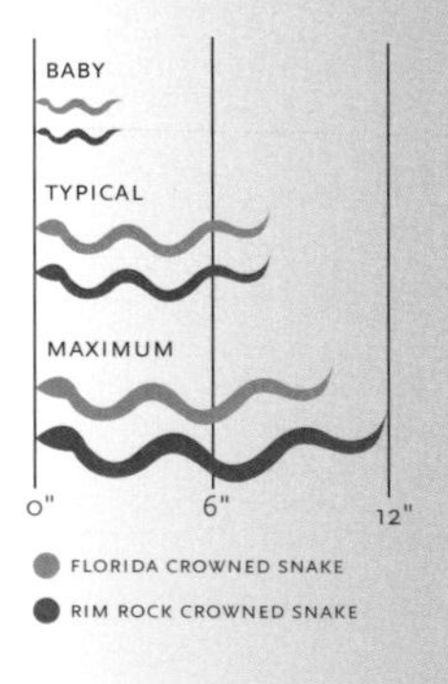

Florida Crowned Snake *Tantilla relicta*
Rim Rock Crowned Snake *Tantilla oolitica*

DESCRIPTION Both species are slender and brown to reddish brown with a black head and a solid white to pinkish belly. The rim rock crowned snake (*T. oolitica*) has dark coloration on the head that extends onto the neck, and sometimes an obscure, light-colored neckband is present. The Florida crowned snake (*T. relicta*) is usually more reddish brown with a dark head and dark neck band. The three subspecies of the Florida crowned snake (peninsula crowned snake, *T. r. relicta*; coastal dunes crowned snake, *T. r. pamlica*; and central Florida crowned snake, *T. r. neilli*) are distinguished primarily by subtly different head and neck patterns.

WHAT DO THE BABIES LOOK LIKE? Baby Florida and rim rock crowned snakes look like their parents.

DISTRIBUTION AND HABITAT Both of these Florida species are most commonly associated with sandy habitats such as pine and scrub oak sandhills, pine flatwoods, and hammocks. The rim rock crowned snake, found only on the southeastern tip of peninsular Florida and the Florida Keys, has the most limited geographic range of any southeastern snake. The Florida crowned snake has also been reported from Valdosta County, Georgia.

BEHAVIOR AND ACTIVITY Both species are presumed to be active during all but the coldest periods in winter but spend most of their time beneath the soil or under rocks and ground litter. They are seldom active aboveground.

FOOD AND FEEDING The Florida crowned snake is known to eat beetle larvae, and may also eat centipedes and snails. Scientists do not know what the rim rock crowned snake eats, but centipedes and other small invertebrates are presumed to constitute a major part of its diet. Like other species of *Tantilla*, the two crowned snakes found exclusively or primarily in Florida have enlarged rear teeth that presumably direct venom into their prey. The rear fangs and toxic saliva may help to subdue prey, but both species are harmless to humans.

Florida crowned snake (coastal dunes subspecies)

REPRODUCTION Scientists know little about reproduction in Florida crowned snakes and essentially nothing about the reproductive biology of rim rock crowned snakes. Both species are egg layers, and their reproduction is likely similar to that of other members of the genus *Tantilla*. The little known about Florida crowned snake reproduction suggests that eggs can be laid anytime from late spring until August.

PREDATORS AND DEFENSE Like other species of crowned snakes, both species are eaten by an array of predators

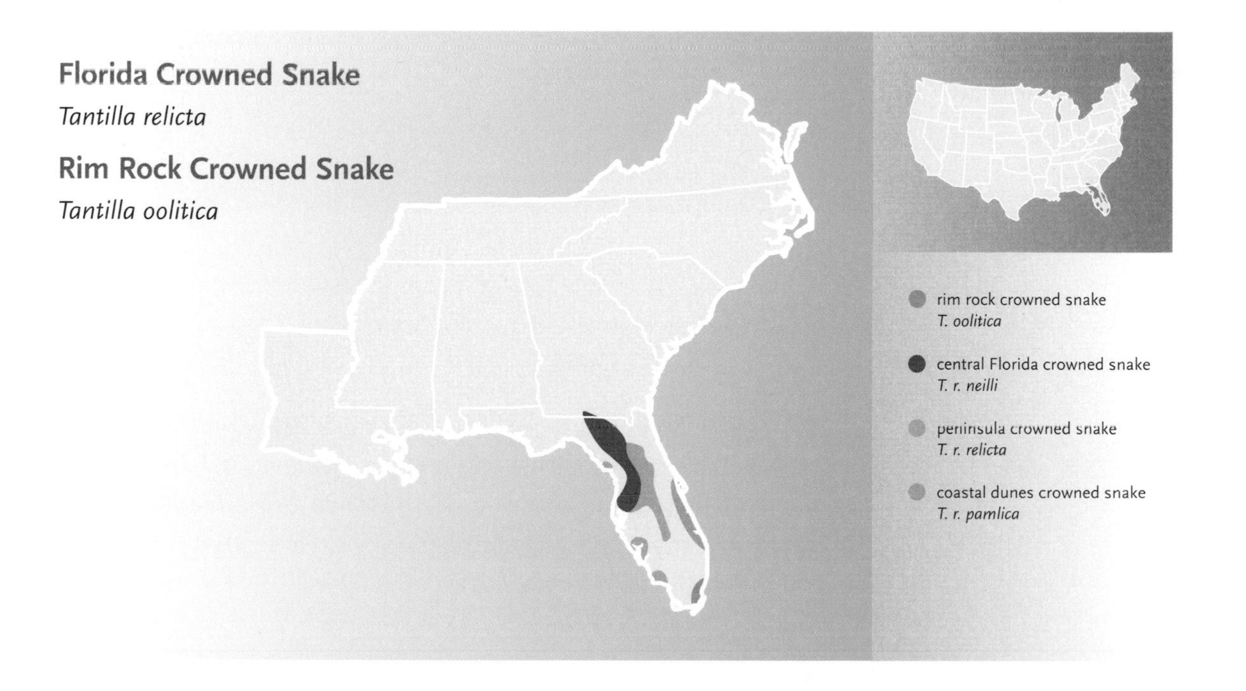

Central Florida crowned snake

Florida crowned snake (peninsula subspecies)

Did you know?

A native scorpion has been suggested to be a natural predator of the rim rock crowned snake.

able to find and capture small snakes underground or beneath ground litter, including other snakes. Neither the rim rock nor the Florida crowned snake will bite when handled.

CONSERVATION Some herpetologists consider the rim rock crowned snake to be threatened because of its limited geographic range and the rapid urban development concentrated in southern Florida. The more widespread Florida crowned snake is apparently in a less precarious situation, but some populations are probably threatened by habitat loss.

How do you identify a flat-headed snake?

SCALES
Smooth

ANAL PLATE
Divided

BODY SHAPE
Slender

BODY PATTERN AND COLOR
Light brown or gray with dark gray head

SIZE

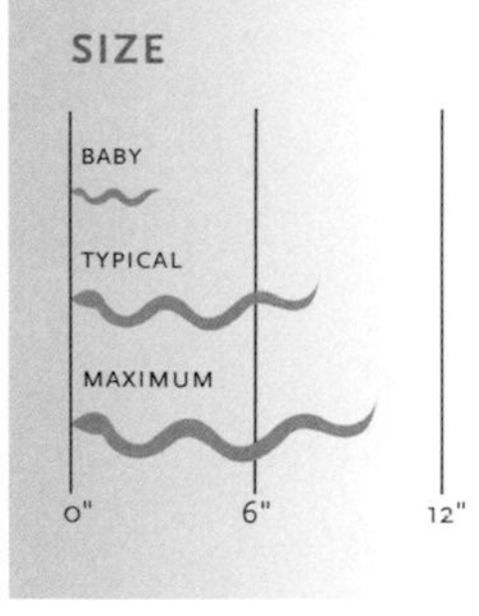

Flat-headed Snake *Tantilla gracilis*

DESCRIPTION Flat-headed snakes are small, slender, brown or grayish snakes with pink bellies. As the name implies, the dark gray head is rather flattened. The head is also somewhat pointed, and the lower jaw is countersunk.

WHAT DO THE BABIES LOOK LIKE? Baby flat-headed snakes look like the adults.

DISTRIBUTION AND HABITAT This species is found in hardwood or mixed hardwood-pine forests, especially in rocky areas, in northwestern Louisiana, often under rocks and frequently associated with limestone habitat.

BEHAVIOR AND ACTIVITY Flat-headed snakes are most active in spring and fall; they hibernate during winter and presumably remain underground during much of the summer. Aboveground activity usually occurs in early evening and at night during warm months.

FOOD AND FEEDING Like many members of the genus *Tantilla*, flat-headed snakes eat centipedes, beetle and other insect larvae, spiders, and slugs. Slightly enlarged, grooved teeth at the back of the mouth direct weak venom into the prey while the snake is biting it. These snakes pose no danger to humans.

Flat-headed snakes are found in northwestern Louisiana.

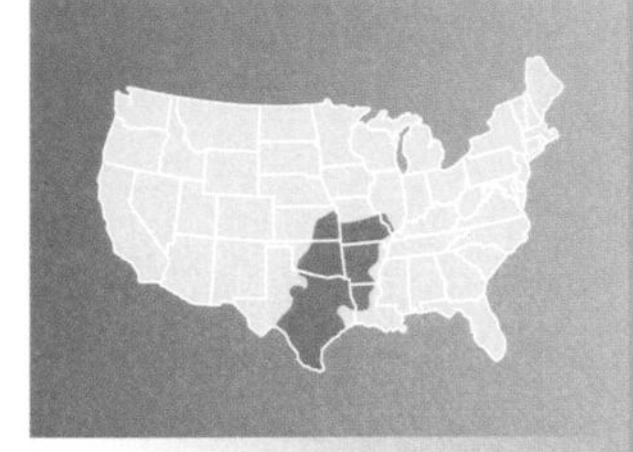

Flat-headed snake

Tantilla gracilis

REPRODUCTION Mating occurs in spring, and two to four eggs are laid in June and early July, usually under rocks or in moist sand. The eggs typically hatch in August and September.

PREDATORS AND DEFENSE Native mammals, birds, and reptiles (snakes and lizards) are natural predators of these little snakes. When disturbed, flat-headed snakes will try to escape by burrowing underground or beneath debris. If captured, they will not bite, although they may release musk.

CONSERVATION This species currently faces no particular conservation threats.

Short-tailed snakes may have orange, yellow, or red on the body.

How do you identify a short-tailed snake?

SCALES
Smooth

ANAL PLATE
Single

BODY SHAPE
Very slender

BODY PATTERN AND COLOR
Primarily gray body with dark blotches on the back and sides sometimes interspersed with light orange or yellow blotches

DISTINCTIVE CHARACTERS
Very short tail

SIZE

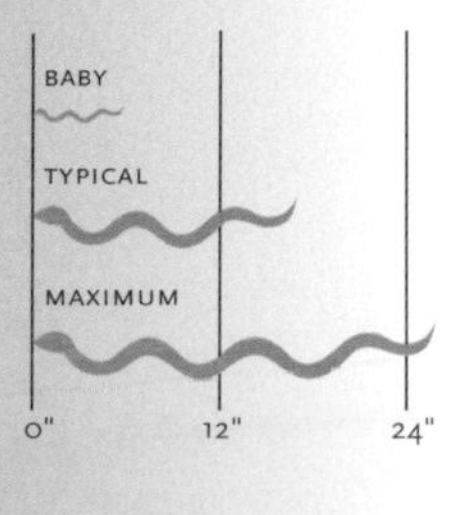

Short-tailed Snake *Stilosoma extenuatum*

DESCRIPTION This very slender snake is gray with a row of dark brown blotches running along the center of the back and a row of smaller dark brown blotches along each side. The gray background color is often interspersed with yellow, orange, or red, and the head is dark. The belly is gray to brown with small white spots.

WHAT DO THE BABIES LOOK LIKE? Baby short-tailed snakes look like the adults.

DISTRIBUTION AND HABITAT The short-tailed snake is known only from about a dozen Florida counties along the sand ridge that runs from the north-central region of the state near the Georgia border to the south-central region above Lake Okeechobee. Because of this snake's secretive behavior and presumed rarity, distribution records are scarce within the known geographic range. The species is confined almost exclusively to sandhill habitats of longleaf pine–turkey oak stands and adjacent upland hardwoods. A few specimens were found in a sphagnum bog near a sandhill habitat.

BEHAVIOR AND ACTIVITY The short-tailed snake apparently spends most of its life underground and is seen so rarely that firm statements about some aspects of its behavior, including daytime activity periods, cannot be made. Most of the active aboveground specimens found were encountered during the day in April or October.

FOOD AND FEEDING The natural diet is unknown. Captive specimens have been known to eat Florida crowned snakes, and presumably they eat this species in the wild as well. They may also prey on other small reptiles such as ringneck snakes, brown snakes, and ground skinks. The short-tailed snake is a constrictor.

REPRODUCTION The number of eggs laid is unknown, as are the details of courtship and mating, where the eggs are laid, and the incubation period.

Short-tailed snake from Gainesville, Florida

PREDATORS AND DEFENSE Predation on this species has not been observed, but presumably snake-eating snakes such as kingsnakes, racers, and coral snakes are natural predators. When captured, short-tailed snakes attempt to escape and may vibrate the tail and bite.

CONSERVATION Short-tailed snakes are so seldom seen that their actual status is difficult to ascertain. They appear to be very rare, but that may merely reflect their extremely secretive habits. Conservationists consider the greatest threat to the short-tailed snake to be loss of longleaf pine–turkey oak habitat in Florida due to commercial development. Florida herpetologists would appreciate receiving news of any sightings.

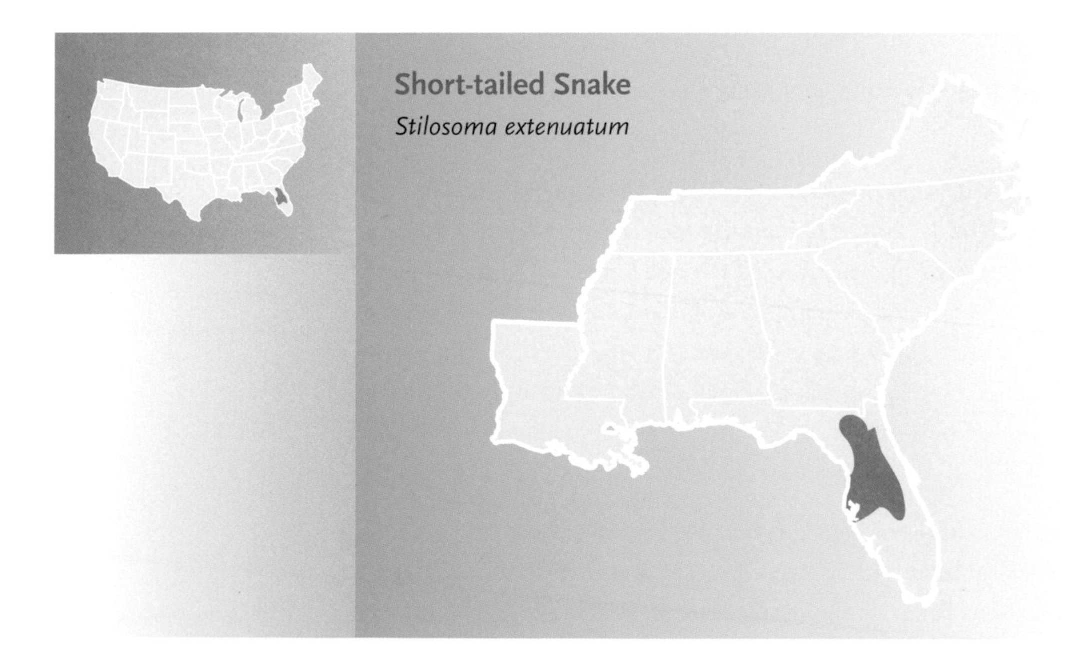

Pine woods snakes are sometimes called yellow-lipped snakes.

How do you identify a pine woods snake?

SCALES
Smooth

ANAL PLATE
Divided

BODY SHAPE
Slender

BODY PATTERN AND COLOR
Solid orange to reddish brown back with whitish belly

DISTINCTIVE CHARACTERS
Dark line passing through eye; lip scales white or yellowish

SIZE

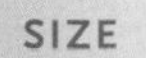

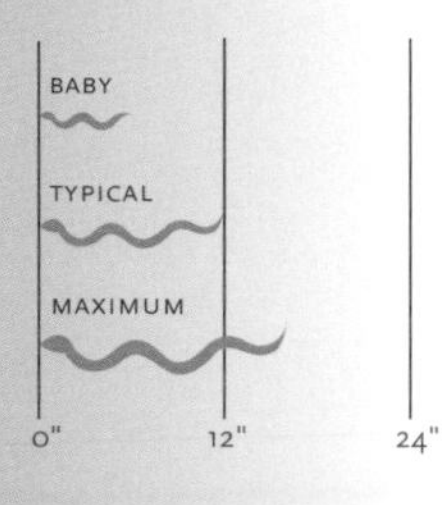

Pine Woods Snake *Rhadinaea flavilata*

DESCRIPTION Pine woods snakes are light to reddish brown or dark orange above with a white to yellow unmarked belly. A faint stripe may be present along the middle of the back. A dark brown stripe starts at the nose and extends through the eye to the corner of the mouth. The chin and lip scales are yellowish, white, or cream colored.

WHAT DO THE BABIES LOOK LIKE? Baby pine woods snakes look like the adults.

OTHER NAMES Pine woods snakes are also known as yellow-lipped snakes.

DISTRIBUTION AND HABITAT Although seldom common, this species inhabits pine and mixed pine-hardwood forests. Individuals have been reported from hardwood hammocks in Florida, from pine areas on coastal islands, and even from inside houses. This species has a patchy geographic distribution, with large gaps apparently occurring between known populations.

BEHAVIOR AND ACTIVITY Because they are found mostly in warm coastal areas, pine woods snakes are active for much of the year, although they may hibernate underground or in logs during cold winters. They are most fre-

Pine woods snakes from some areas have a reddish color.

Did you know?

Many small southeastern snakes that are considered harmless to humans have enlarged teeth in the back of the mouth (rear-fanged) that facilitate the transfer of toxic saliva into their prey.

quently encountered in the spring, but they are seldom seen aboveground and their daily activity patterns are unknown.

FOOD AND FEEDING Very little is known of the natural diet. Captive specimens prefer small frogs, salamanders, and an occasional small lizard. Enlarged teeth in the back of the mouth direct toxic saliva into the prey. The bite subdues the small prey animals but is of no consequence to humans.

REPRODUCTION Little information exists regarding reproduction; mating apparently occurs in spring, and one to four (most often two or three) eggs are laid during summer. Some females may lay two clutches of eggs during a single year. The babies hatch 6–8 weeks later.

Pine Woods Snake

Rhadinaea flavilata

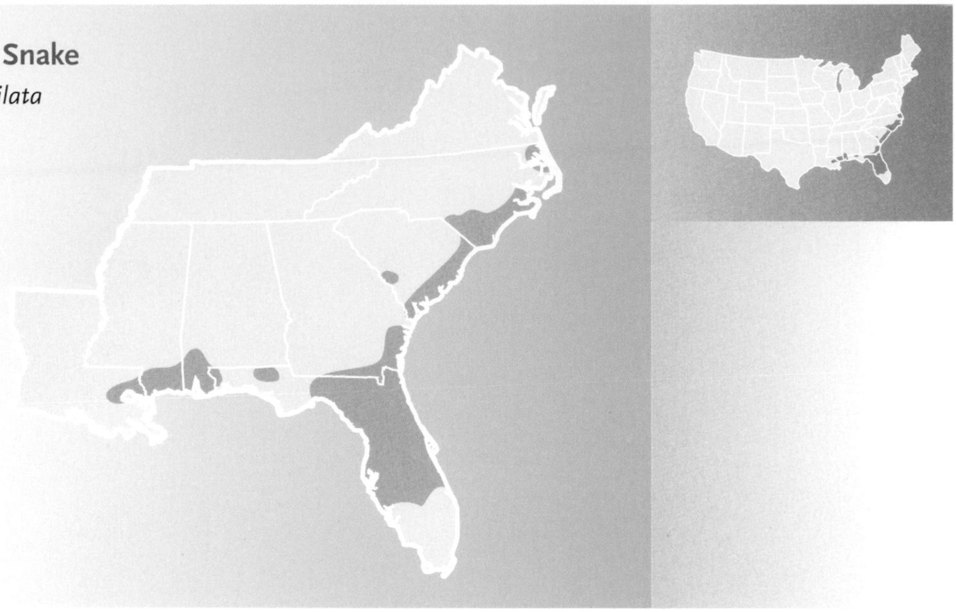

PREDATORS AND DEFENSE Terrestrial snakes such as racers and kingsnakes have been reported to eat pine woods snakes, but other carnivorous pine forest animals such as shrews, birds, and toads are also potential predators. Pine woods snakes do not bite when picked up by humans but may release a bad-smelling musk.

CONSERVATION Conservation threats to this species are difficult to confirm because its secretive nature and scattered distribution make the documentation of declines or disappearances difficult. Habitat destruction, degradation, and fragmentation due to human activities are likely the greatest threats these snakes face.

Pine woods snake from Tampa, Florida

Western worm snakes are black above.

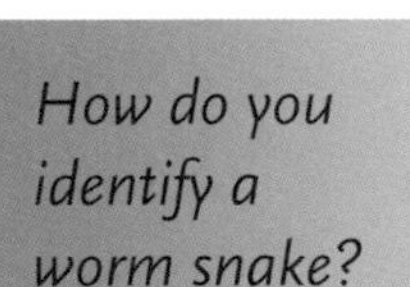

SCALES
Smooth

ANAL PLATE
Divided

BODY SHAPE
Slender; very small head

BODY PATTERN AND COLOR
Solid light brown (eastern) or black (western) above; pinkish belly

DISTINCTIVE CHARACTERS
Spine on tip of tail; tiny eyes

SIZE

BABY

TYPICAL

MAXIMUM

0" 12" 24"

EASTERN WORM SNAKE

WESTERN WORM SNAKE

Eastern Worm Snake — *Carphophis amoenus*
Western Worm Snake — *Carphophis vermis*

DESCRIPTION Worm snakes are either brown, dark gray, or black above. The belly is either white or pink, and the light belly coloration extends slightly up onto the sides. The head is tiny and pointed, and there is a sharp point or spine on the tip of the tail. The eastern worm snake (*C. amoenus*) is usually more brownish than the western worm snake (*C. vermis*). The two subspecies of the eastern worm snake (eastern worm snake, *C. a. amoenus;* and midwestern worm snake, *C. a. helenae*) are difficult to tell apart without counting the small head scales.

WHAT DO THE BABIES LOOK LIKE? Babies resemble the adults, but young eastern worm snakes may be slightly darker than their parents.

DISTRIBUTION AND HABITAT Worm snakes are found in portions of all southeastern states except Florida, typically in cool, moist hardwood forests. They are extremely abundant in the Piedmont of the Southeast, especially along rocky, wooded hillsides.

BEHAVIOR AND ACTIVITY Worm snakes hibernate in most areas but are active from early spring to late fall. They remain underground or beneath rocks and logs for most of their lives. Their occasional aboveground forays take place primarily at night during the warmest months.

FOOD AND FEEDING Worm snakes feed almost exclusively on earthworms, although they may rarely take soft-bodied insect larvae or slugs, and one western worm snake was reported to have eaten a small ringneck snake. The worm snake grasps an earthworm anywhere along its body, moves its mouth to one end or the other, and begins swallowing.

REPRODUCTION Reproduction is best known in western worm snakes, which breed in the spring and again in the fall. Some evidence suggests that eastern worm snakes may do this as well. Both species lay 1–12 eggs (usually 4–5) under rocks or in rotting logs in June or July, and the babies hatch in late summer or early fall.

PREDATORS AND DEFENSE Because they are among the smallest of the southeastern snakes, worm snakes are vulnerable to a variety of woodland snakes, birds, and mammals. When captured, worm snakes wriggle vigorously and are unique among the small southeastern snakes in pressing the spine of the tail against the captor, but never with enough force to penetrate the skin. They release a strong-smelling fluid from cloacal glands (anal region) when disturbed, but almost never bite humans.

Worm snakes have small heads and tiny eyes.

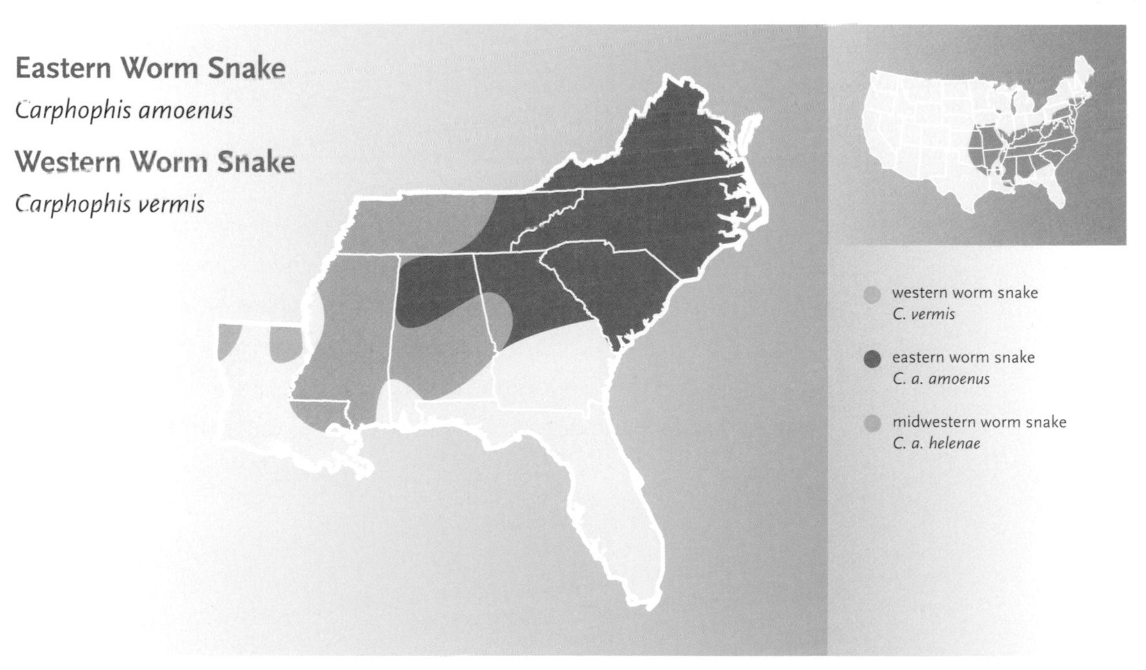

Worm snakes have pointed tails.

Most worm snakes have pink bellies.

CONSERVATION The loss of woodland habitat due to human development is a potential threat to both species, and insecticides have been implicated in the deaths of eastern worm snakes. Worm snakes have been reported to have died in floods of low-lying habitats, presumably because of their tendency to remain underground. Thus, reservoirs or other damming projects that flood wooded areas could be detrimental to worm snake populations.

Northern ringneck snake with complete neck ring

Ringneck Snake

Diadophis punctatus

DESCRIPTION These very slender snakes are usually gray to nearly black above with a yellow, orange, or red belly and a light ring around the neck. In some subspecies, the belly usually has a row of dark spots down the center. Four subspecies occur in the Southeast. The southern ringneck snake (*D. p. punctatus*) has a neck band that is broken in the middle and a row of black spots running along the middle of its belly. The northern ringneck snake (*D. p. edwardsi*) has a complete neck ring and no large spots on the belly. The Mississippi ringneck snake (*D. p. stictogenys*) has a very narrow, sometimes broken neck band and irregular spots along the center of the belly. The Key ringneck snake (*D. p. acricus*), known only from the Florida Keys, has almost no neck ring.

WHAT DO THE BABIES LOOK LIKE? Baby ringneck snakes look like the adults.

DISTRIBUTION AND HABITAT Ringneck snakes range throughout the southeastern states except for portions of southwestern Louisiana and are commonly associated with wooded habitats such as mixed hardwood and pine forests in mountainous areas, swamp margins, and floodplains. The species is noted for requiring ground cover such as rocks, logs, or ground litter for hiding.

How do you identify a ringneck snake?

SCALES
Smooth

ANAL PLATE
Divided

BODY SHAPE
Slender

BODY PATTERN AND COLOR
Gray or black, usually with light ring around neck; belly yellow to red, sometimes with small black dots

SIZE

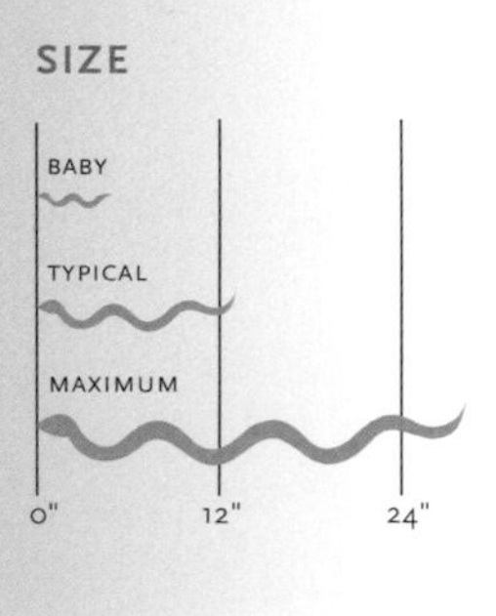

Ringneck snake from Kennesaw Mountain near Atlanta

BEHAVIOR AND ACTIVITY Ringneck snakes are active throughout the year in warm regions, such as most of Florida, and during the warm months elsewhere. They hibernate in areas of the Southeast with cold winters. These small snakes are noted for relatively long-distance travel—more than a mile in some instances and frequently more than 200 feet—but they seem more averse than many other snake species to crossing roads, particularly those with moderate to heavy traffic. Activity aboveground is primarily at night.

Mississippi ringneck snake

FOOD AND FEEDING The diet includes a variety of small animals such as earthworms, salamanders, small frogs, small snakes and lizards, and insect larvae. They may paralyze prey by chewing on it with their enlarged posterior teeth so that their toxic saliva can enter the lacerations. Although ringneck snakes rarely bite humans and are generally considered harmless, bites have sometimes been reported to cause a burning sensation.

REPRODUCTION Mating occurs in spring and again in fall. Females store the sperm from fall matings over the winter, and fertilization occurs in the spring. When courting, the male ringneck rubs his mouth along the neck of the female and bites her on her neck ring. Ringneck snakes lay two to seven eggs in moist, protected sites (e.g., in rotting logs and under rocks) during summer. Good nesting sites are sometimes used by more than one female. The eggs hatch in about 7–8 weeks.

PREDATORS AND DEFENSE The number of known natural predators is high because of the extensive research that has been done on this abundant and widespread species. No fewer than eight species of terrestrial snakes have been reported to eat ringneck snakes, as have five species of predatory birds, six na-

Ringneck Snake

Diadophis punctatus

tive mammals, bullfrogs, and toads. The ringneck snake's first response to capture is to attempt to escape. When exposed while hiding beneath a log, the snake may momentarily flip over, exposing the bright yellow, orange, or red belly, and then turn back over and crawl into dark soil or rotting wood. Whether intentional or not, the maneuver can momentarily distract a predator (or a person) and allow the dark-bodied snake to escape. Ringneck snakes also produce a musk that some people find extremely disagreeable. Some captured or disturbed individuals become motionless, as if dead; some may hide the head beneath the body; and some display the underside of the brightly colored tail, which may appear threatening to some small predators. These snakes seldom bite people.

CONSERVATION Ringneck snakes are extremely abundant in many areas, but like other species that require woodland habitats with a broad array of terrestrial prey, they are vulnerable to environmental alterations that destroy natural habitats or introduce pesticides into the ecosystem.

Northern ringneck snake with complete neck ring and southern ringneck snake with interrupted neck ring

Key ringneck snake from Marathon, Florida

MID-SIZED TERRESTRIAL SNAKES

Scarlet Snake

Cemophora coccinea

DESCRIPTION Scarlet snakes appear to have rings or bands, although the red, yellow or white, and black blotches that cover most of the back do not actually encircle the body. The belly is immaculate white. The pointed nose is always red followed by a narrow black band. The two subspecies that occur in the Southeast (northern scarlet snake, *C. c. copei*; and Florida scarlet snake, *C. c. coccinea*) are distinguished by minor differences in scale numbers and spacing of the banding pattern.

WHAT DO THE BABIES LOOK LIKE? Baby scarlet snakes look like the adults.

DISTRIBUTION AND HABITAT The geographic range includes almost all areas outside of mountainous habitat within every southeastern state except the central and southern portions of Louisiana. Scarlet snakes are found throughout much of the Coastal Plain and are more patchily distributed in the Piedmont of most states. They are associated mostly

Florida scarlet snake from Alachua County

How do you identify a scarlet snake?

SCALES
Smooth

ANAL PLATE
Single

BODY SHAPE
Slender

BODY PATTERN AND COLOR
Red, yellow or white, and black blotches or saddles on back; belly white

DISTINCTIVE CHARACTERS
Pointed, red nose

SIZE

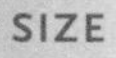

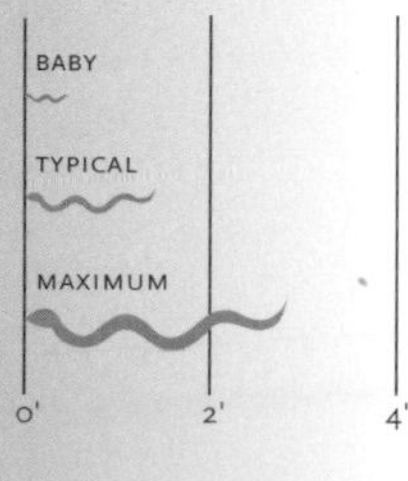

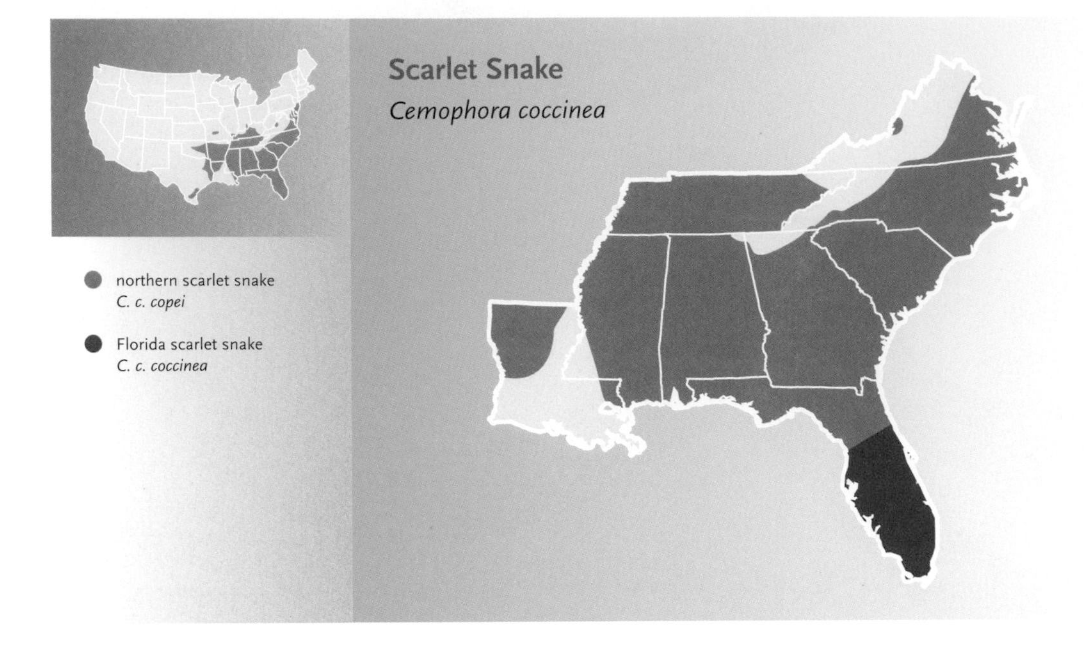

with pine, hardwood, and mixed hardwood-pine forests. Generally, they prefer open forests with sandy soil and lots of ground litter and organic debris under which they can retreat. They are considered fossorial (i.e., they dig or burrow) and use the pointed nose to burrow through loose soil.

Did you know?

Young scarlet snakes are so tiny that southern toads have been reported to feed on them.

BEHAVIOR AND ACTIVITY Scarlet snakes are active aboveground almost exclusively at night and only during the warmer months. They are often found crossing roads at night or beneath logs, pine straw, and other ground cover.

FOOD AND FEEDING Scarlet snakes feed primarily on the eggs of other reptiles (snakes, lizards, and turtles), but they also use constriction to subdue and eat small lizards, snakes, and frogs. The enlarged, knifelike posterior teeth are used to slit open large reptile eggs. The snake feeds either by squeezing the egg to expel the contents or by sticking its head into the egg. Small reptile eggs may be eaten whole.

REPRODUCTION Very little is known about the reproductive biology of scarlet snakes. They are egg-layers that apparently mate in the spring and lay their elongated eggs during the summer in underground burrows, under rocks, or in similar places. Clutches average about five eggs (but can range from two to nine). The eggs hatch after about 2 months, usually during late summer or fall.

PREDATORS AND DEFENSE The known natural predators are animals of sandhill regions such as coral snakes, other snake-eating snakes, predatory

Scarlet snakes have red heads and pointed noses; the solid white belly distinguishes scarlet snakes from scarlet kingsnakes.

birds, and mammals. Scarlet snakes seldom bite when captured although they sometimes release a mild-smelling musk.

CONSERVATION Two of the greatest threats to scarlet snakes result from fragmentation of their habitat by roads (they suffer extensive on-road mortality in areas of heavy nighttime traffic) and commercial development and associated destruction of their natural habitats.

Northern scarlet snake from North Carolina

Rough green snakes spend most of their time aboveground in green vegetation.

How do you identify a rough green snake?

SCALES
Keeled

ANAL PLATE
Divided

BODY SHAPE
Extremely slender

BODY PATTERN AND COLOR
Solid green above; typically pale yellow below

DISTINCTIVE CHARACTERS
Body color turns blue when snake is dead

SIZE

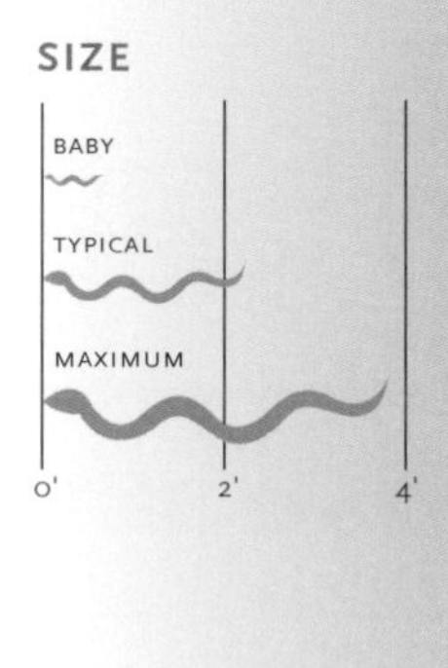

Rough Green Snake *Opheodrys aestivus*

DESCRIPTION The rough green snake is almost vinelike in appearance and is always solid green above. The belly is usually yellowish or yellowish white. The head is somewhat distinct, and the eyes are rather large.

WHAT DO THE BABIES LOOK LIKE? Baby green snakes are grayish green rather than bright green like the adults.

DISTRIBUTION AND HABITAT Rough green snakes are found throughout every southeastern state except for a few mountainous areas in Virginia. They like to climb and are usually found in bushes and vines along waterways, roads, forests, and swamps. They are often found near water.

BEHAVIOR AND ACTIVITY Rough green snakes are active almost exclusively during the daytime hours throughout the warmer months. They commonly hibernate underground or in stumps or logs during winter, and have even been found in ant mounds.

FOOD AND FEEDING The diet consists mostly of insects and insect larvae, but millipedes, spiders, land snails, and small treefrogs are also eaten. Green snakes usually grab and swallow their prey alive but may chew on very active animals to subdue them before swallowing.

Rough Green Snake

Opheodrys aestivus

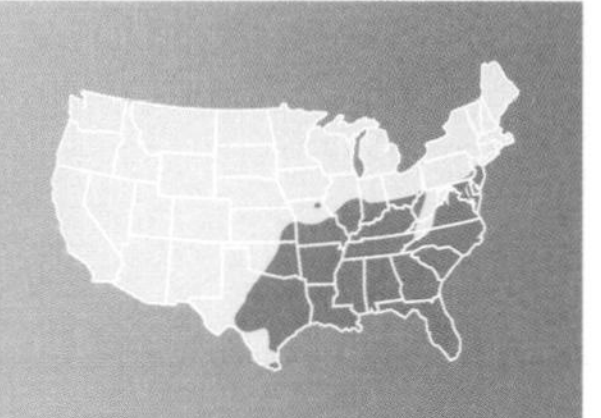

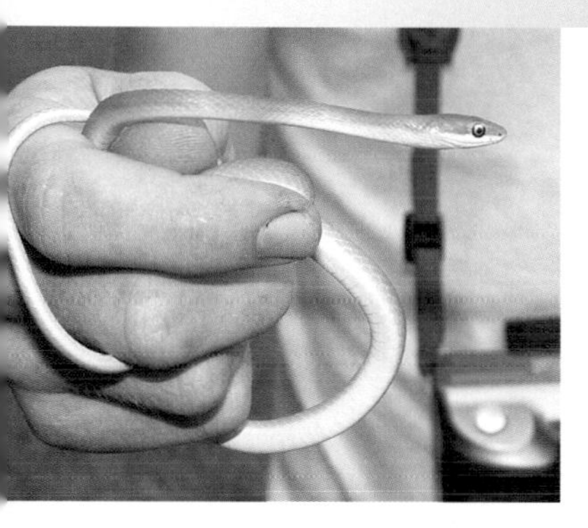

Rough green snakes are solid green above and yellow or whitish below.

REPRODUCTION Mating occurs primarily in the spring, but they may also mate in the fall. The female lays 3–12 eggs in rotting or hollow logs, under rocks, or in other concealed moist places, usually during the summer. Several females may lay their eggs in the same location.

PREDATORS AND DEFENSE Birds, spiders, and other snakes have been documented to eat rough green snakes. The bright green color is an ideal camouflage in vines, shrubs, and bushes, where the snake may remain completely motionless while a person or predator is near. When captured, green snakes may open their mouths but seldom actually bite.

CONSERVATION Because they are often associated with the margins of aquatic areas, development that destroys shrubbery or other vegetation around streams and wetlands may reduce available habitat for rough green snakes. Because they eat insects and spiders, they may be particularly susceptible to pesticides. Green snakes are often killed on highways adjacent to wetland or woodland habitats.

Rough green snakes hunt by sight for insects and spiders.

Smooth green snakes are typically found on the ground, often near streams or wetlands.

How do you identify a smooth green snake?

SCALES
Smooth

ANAL PLATE
Divided

BODY SHAPE
Slender

BODY PATTERN AND COLOR
Solid green above; white below

DISTINCTIVE CHARACTERS
Body color turns blue when snake is dead

SIZE

BABY
TYPICAL
MAXIMUM
0"
12"
24"

Smooth Green Snake *Opheodrys vernalis*

DESCRIPTION The body is solid green above with a solid white or pale yellowish belly. Smooth green snakes are somewhat more robust than rough green snakes.

WHAT DO THE BABIES LOOK LIKE? Baby smooth green snakes are more grayish or olive than the bright green adults.

DISTRIBUTION AND HABITAT The geographic range in the Southeast is limited to mountainous areas of northwestern Virginia, where they are generally found on the ground in moist meadows or grassy areas and at the edges of bogs or small streams, often beneath rocks or other debris.

BEHAVIOR AND ACTIVITY Smooth green snakes are active during the day throughout the warmer months. In winter they typically hibernate underground or in stumps or logs, and sometimes are found in ant mounds.

FOOD AND FEEDING Smooth green snakes eat insects and their larvae, spiders, centipedes, millipedes, worms, small salamanders, and small crayfish. They are not constrictors, but simply grab and swallow their prey.

REPRODUCTION Smooth green snakes mate in the spring and possibly in the fall, and lay two to seven eggs in rotting logs, under rocks, or in other

Smooth green snake

concealed moist places. The eggs go through extensive development while still inside the female and thus hatch about 2 weeks or less after they are laid.

PREDATORS AND DEFENSE Snake-eating snakes, hawks, and spiders have been documented to eat smooth green snakes, which depend on their bright green color for camouflage in grassy areas. Smooth green snakes typically do not bite when picked up.

CONSERVATION Because they are commonly associated with the margins of aquatic areas, development that destroys habitats around small streams and wetlands may eliminate some populations. They are also frequent victims of road mortality on highways near their habitats.

SCIENTIFIC NOMENCLATURE Some herpetologists place the smooth green snake in the genus *Liochlorophis* rather than in *Opheodrys* with the rough green snake.

Did you know?

The smooth green snake is the only southeastern snake species that is found only in Virginia.

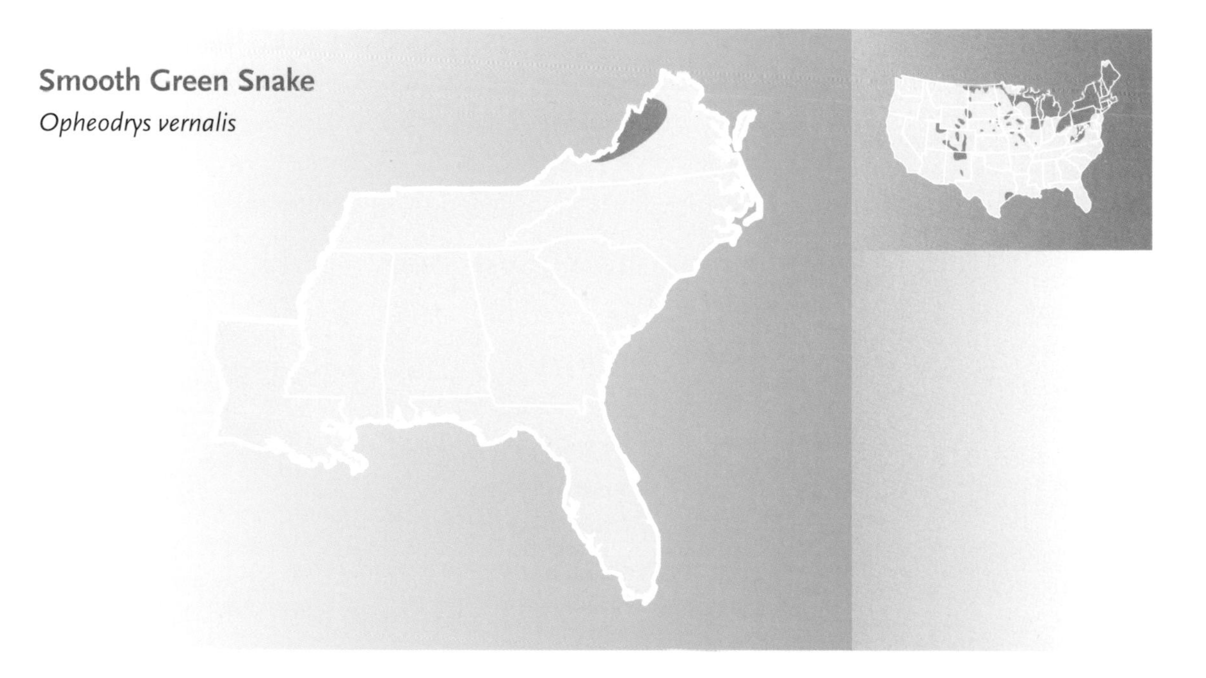

Eastern garter snakes typically have three distinct yellow stripes.

How do you identify an eastern garter snake?

SCALES
Keeled

ANAL PLATE
Single

BODY SHAPE
Moderately stocky

BODY PATTERN AND COLOR
Highly variable but typically with a dark body and either three stripes (usually yellowish) or a checkered appearance above and a solid pale yellow belly

DISTINCTIVE CHARACTERS
A dark vertical line borders each yellow lip scale

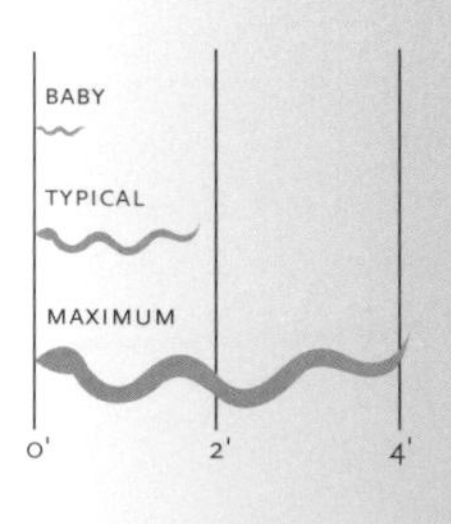

Eastern Garter Snake *Thamnophis sirtalis*

DESCRIPTION Eastern garter snakes (*T. s. sirtalis*) are quite variable in appearance, but most have three yellow, brownish, or greenish stripes on a dark body. Some specimens have a vague checkerboard pattern that may obscure the stripes to some degree. The belly is usually a solid color, ranging from pale yellow to whitish to greenish with a black mark on the outer edge of each belly scale. The lip scales are yellowish or white with distinct vertical dark markings between them. The blue-striped garter snake (*T. s. similis*), the only other subspecies in the Southeast, has a dark brown body with a pale stripe down the center and a light blue stripe on each side.

WHAT DO THE BABIES LOOK LIKE? Baby garter snakes look like the adults.

Baby eastern garter snake

DISTRIBUTION AND HABITAT Eastern garter snakes are found throughout every southeastern state except in some areas in western and southern Louisiana. Suitable habitats are generally near water and include the edges of drainage ditches, streams, swamps, ponds, and wet meadows and pastures. They prefer open, grassy

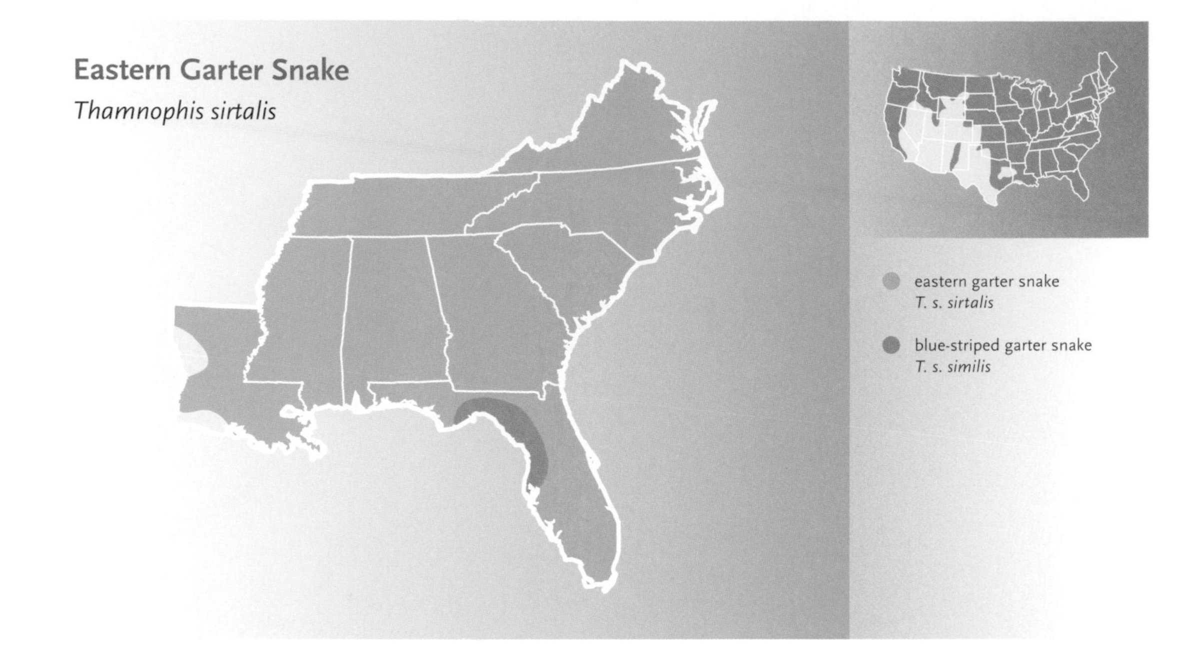

habitats but are occasionally found in wooded areas.

BEHAVIOR AND ACTIVITY Eastern garter snakes are noted for their tolerance of cool temperatures and are often active during the winter on warm or sunny days basking on bushes, rocks, or grassy areas along streams or ponds. Many studies have shown that garter snakes will travel distances of more than a mile during migration from feeding habitats to communal hibernation sites, but in the Southeast they may hibernate alone. They are active primarily during the day but occasionally forage at night during the summer.

The lip scales of eastern garter snakes are separated by black vertical lines.

FOOD AND FEEDING The primary food is amphibians, but the diet may include many kinds of invertebrates, fish, small snakes, baby birds, mice, and shrews. Prey is detected primarily by scent. After capturing active prey such as mice, eastern garter snakes may chew vigorously, presumably so that their mildly toxic saliva will help to subdue their prey.

Blue-striped garter snake from Taylor County, Florida

An eastern garter snake from Aiken, South Carolina

REPRODUCTION Eastern garter snakes mate primarily in the spring, although fall matings have been observed. Females that are ready to mate release pheromones that attract males, sometimes many males, which may try to mate simultaneously with one female. Females can store sperm from previous matings for many months, and thus the offspring in a single litter may have multiple fathers. During development, the embryos are nourished through a placenta-like structure. Young are born during the summer or fall, and litter size ranges from 1 to 101 (usually = 20–30).

PREDATORS AND DEFENSE Their long-distance movements, activity during most months of the year, and association with both terrestrial and aquatic habitats expose garter snakes to a broad spectrum of predators, including snakes, turtles, birds, mammals, spiders, frogs, and fish. A garter snake's first response when threatened is to try to escape. Some feign death and/or hide the head beneath the body. Captured individuals may expand the head and body to appear larger and usually deliver open-mouthed strikes, eventually biting and chewing on the captor. They release a sweet-smelling,

Eastern garter snakes often bask in the sun during cool weather.

Some eastern garter snakes have a checkered appearance.

somewhat nauseating musk when first captured. The saliva of garter snakes is mildly venomous and can cause swelling and itching in some people.

CONSERVATION In California, a garter snake subspecies (the San Francisco garter snake, *T. s. tetrataenia*) is federally protected because so much of its habitat has been lost, and many populations in the Southeast are being lost as a result of commercial development as well. Garter snakes are frequent victims of highway mortality in many regions of the United States and Canada.

Typical appearance of the eastern ribbon snake

How do you identify a ribbon snake?

SCALES
Keeled

ANAL PLATE
Single

BODY SHAPE
Very slender

BODY PATTERN AND COLOR
Typically with three yellow stripes above and a pale whitish or yellowish belly

DISTINCTIVE CHARACTERS
Tail proportionately longer than that of most other snakes

SIZE

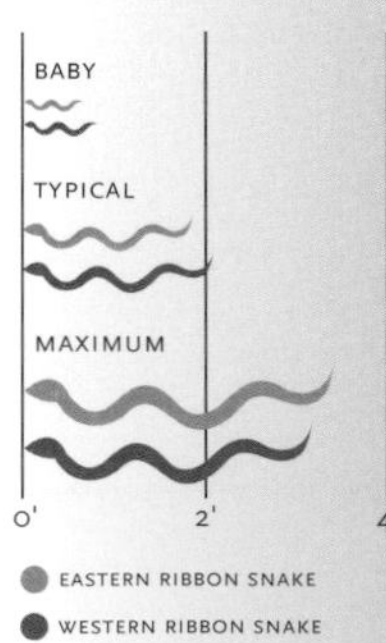

Eastern Ribbon Snake — *Thamnophis sauritus*
Western Ribbon Snake — *Thamnophis proximus*

DESCRIPTION Ribbon snakes characteristically have three light, usually yellow, stripes down the length of the dark body; the belly can be white or yellow. The lip scales are yellowish or white with no vertical dark markings between them. Five subspecies of the two species are recognized in the Southeast, including the eastern ribbon snake (*T. s. sauritus*), which matches the general description of the species. The stripe down the back of the peninsula ribbon snake (*T. s. sackenii*) is either absent, partially present, or less distinct than the side stripes. The blue-striped ribbon snake (*T. s. nitae*) has blue stripes on the sides, and the middle stripe is absent or poorly defined. The western ribbon snake (*T. p. proximus*) typically has distinct yellow or greenish yellow stripes, and the Gulf Coast ribbon snake (*T. p. orarius*) is characterized by a wide, gold central stripe. Intergradation occurs between the two subspecies of western ribbon snakes (see map).

WHAT DO THE BABIES LOOK LIKE? Baby ribbon snakes are identical to the adults.

The lip scales of ribbon snakes do not have black vertical lines like garter snakes do.

Peninsula ribbon snake from Alachua County, Florida

DISTRIBUTION AND HABITAT One or both species are found throughout all or most of each southeastern state except Tennessee, where the western ribbon snake is limited to the western edge of the state. Both species are more aquatic than their close relative the eastern garter snake, and are often abundant along the edges of permanent and semipermanent aquatic areas such as ponds, marshes, swamps, streams, and rivers. They frequently climb into low bushes overhanging or near the water.

BEHAVIOR AND ACTIVITY Ribbon snakes may be found at any season of the year if daily temperatures are warm or sunny conditions allow individuals to bask. They are active almost exclusively during the day but may venture out in early evening or at night on hot days.

Western ribbon snake

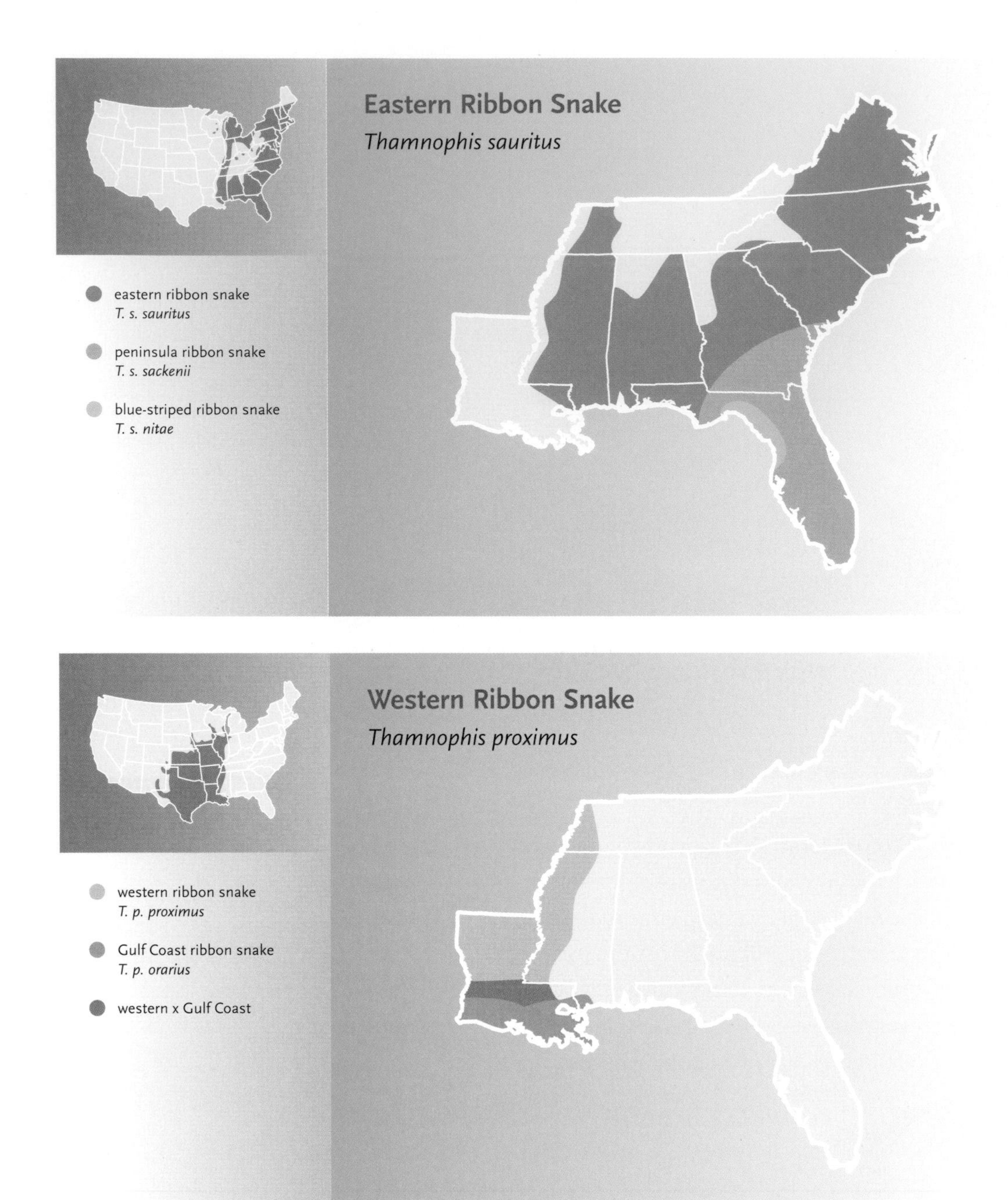
Eastern Ribbon Snake
Thamnophis sauritus
eastern ribbon snake
T. s. sauritus
peninsula ribbon snake
T. s. sackenii
blue-striped ribbon snake
T. s. nitae
Western Ribbon Snake
Thamnophis proximus
western ribbon snake
T. p. proximus
Gulf Coast ribbon snake
T. p. orarius
western x Gulf Coast

FOOD AND FEEDING Ribbon snakes eat primarily amphibians, especially frogs and toads, which they usually swallow rear end first. They also feed on small fish, which they capture by swimming open-mouthed through the water. They detect prey using both vision and smell, and swallow prey while it is still alive.

REPRODUCTION Ribbon snakes mate in spring, and males apparently find females by following their scent trails. The females give birth in the summer or early fall to broods usually ranging from 10 to 15 young. The largest broods recorded are 36 for the western ribbon snake and 26 for the eastern species.

PREDATORS AND DEFENSE Common predators include other snakes (especially kingsnakes and cottonmouths), wading birds, wetland mammals, bullfrogs, and fish. When encountered, ribbon snakes always try to escape either on land or into the water. When captured they release a sickeningly sweet musk and sometimes bite.

CONSERVATION Because of their dependence on wetlands, ribbon snake populations may be affected by the destruction of aquatic habitats and their margins.

Some peninsula ribbon snakes lack the stripe down the back.

Blue-striped ribbon snake from Taylor County, Florida

Body color patterns of eastern hognose snakes are highly variable.

How do you identify an eastern hognose snake?

SCALES
Keeled

ANAL PLATE
Divided

BODY SHAPE
Heavy bodied

BODY PATTERN AND COLOR
Highly variable; sometimes blotched with different colors, including red, yellow, orange, or gray; sometimes solid black

DISTINCTIVE CHARACTERS
Prominent, pointed nose; extraordinary threat display

SIZE

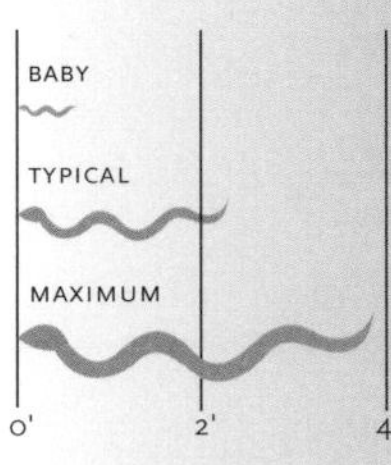

Eastern Hognose Snake *Heterodon platirhinos*

DESCRIPTION Eastern hognose snakes usually have a blotched pattern with color combinations that can vary greatly within a local population or even within the same clutch. The colors on the back and sides can include shades of red, orange, yellow, gray, olive, brown, or black. Some individuals become solid black or dark gray above as adults, and this color pattern is prevalent in some areas of the Southeast. The belly is also variable, ranging from pale gray to dark and sometimes having lighter or darker blotches. The belly is noticeably darker than the undersurface of the tail. The tip of the nose is pointed. The characteristic bluffing and death-feigning displays are so dramatic that they should be considered part of the general description of the species.

WHAT DO THE BABIES LOOK LIKE? Baby hognose snakes, even those that will turn solid black or gray as adults, always have a blotched pattern.

OTHER NAMES Hognose snakes are commonly called spreading adders or puff adders in many parts of the Southeast.

DISTRIBUTION AND HABITAT Eastern hognose snakes occur throughout every southeastern state except for extreme southeastern Louisiana in a wide variety of habitats, including abandoned agricultural fields, open pine

Eastern Hognose Snake

Heterodon platirhinos

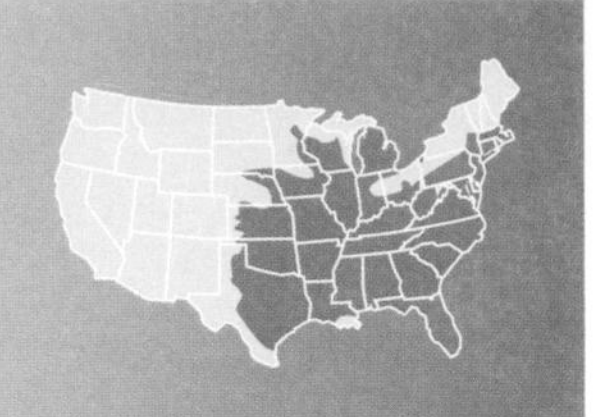

The red color phase of the eastern hognose snake

forests, and rocky forested hillsides. They usually are associated with loose soil, and if living in forests usually stay near the edge where they can venture into more open areas. Hognose snakes characteristically burrow into sandy soils, using the flat, upturned nose as a shovel to push into the soil.

BEHAVIOR AND ACTIVITY Activity in the Southeast generally begins early in the spring and ends late in the fall. Recently born juveniles are commonly found into October, and adults may bask on warm days during the winter months in the warmest parts of the range. Eastern hognose snakes are found active aboveground only during the daytime.

FOOD AND FEEDING Toads along with an occasional frog or salamander make up most of the diet. A few records exist of wild-caught specimens having eaten small mud turtles, lizards, small snakes, birds, and small mammals as well as a variety of invertebrate prey. Toads inflate their bodies as a defense mechanism when captured by snakes, but eastern hognose snakes use their enlarged posterior teeth to deflate the toads and make them easier to swallow. The skin toxins of toads are neutralized by enzymes in the snake's digestive system.

Toads are the most common prey of eastern hognose snakes.

Many adult eastern hognose snakes are black.

Eastern hognose snakes often lay more than two dozen eggs, which hatch in 6–7 weeks.

REPRODUCTION Breeding occurs primarily in the spring but also may occur during the fall. Eggs are laid anytime from May through August, and clutches typically number about 20–25 eggs (range = 4–61). Eggs are usually laid under rocks or in loose soil. Some nests have been found in sawdust piles. The eggs hatch in about 6–7 weeks.

PREDATORS AND DEFENSE Natural predators of eastern hognose snakes are numerous, in part because the snakes are active aboveground in the daytime and occur in a wide array of habitats. Predators include other snakes, birds of prey, and carnivorous mammals. Hognose snakes have a distinctive repertoire of responses to threats from potential predators and humans. When initially threatened, the snake lifts the front of its body off the ground and spreads its neck like a cobra. Some individuals hiss, open the mouth, and move the head forward in a striking motion, but without actually trying to bite. If this formidable display does not intimidate or discourage the attacker, the snake will appear to go into convulsions and roll

Defensive display of the eastern hognose snake; note the tiny rear fangs in the upper jaw.

The belly of eastern hognose snakes is lighter than the underside of the tail. This hognose snake is playing dead.

over on its back as if dead, often disgorging a recently eaten meal as part of the display. The mouth is opened with the tongue hanging out, and capillaries in the mouth may rupture, producing a large amount of blood to add to the effect. The snake remains on its back during the death-feigning display and will actually roll back over if it is righted. Eastern hognose snakes will not intentionally bite people, although injuries may occur accidentally when a person catches a finger on one of the rear fangs when the snake's mouth is held open. The saliva is mildly venomous and sometimes causes an allergic reaction on the rare occasions when someone is bitten.

CONSERVATION Hognose snakes commonly travel overland and frequently become highway mortality victims. Although they are terrestrial, their dependence on toads as prey makes the protection of small wetlands critical for this species.

The sharply upturned snout is characteristic of the southern hognose snake.

How do you identify a southern hognose snake?

SCALES
Keeled

ANAL PLATE
Divided

BODY SHAPE
Very stout

BODY PATTERN AND COLOR
Light brown or sometimes reddish body; dark brown blotches down the back and smaller blotches on the sides

DISTINCTIVE CHARACTERS
Sharply upturned nose

SIZE

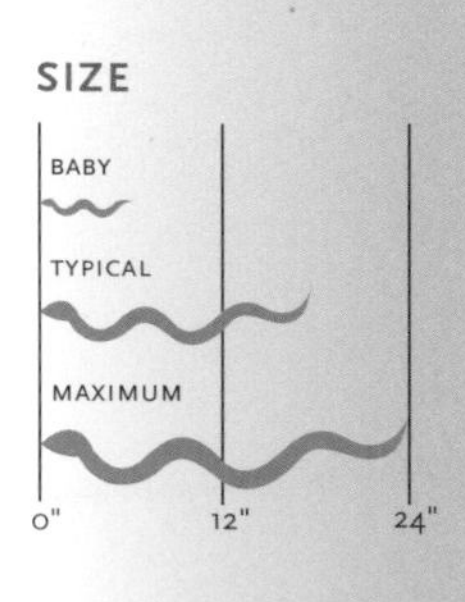

Southern Hognose Snake *Heterodon simus*

DESCRIPTION This is the stoutest snake in the Southeast; adults weigh more than twice as much as many other species at a given length. The end of the nose is turned upward at a sharp angle. The body is usually light brown or reddish brown with dark brown blotches on the back and smaller ones on the sides, and the belly is light brown or gray. Southern hognose snakes are never solid black like eastern hognose snakes. The nose of southern hognose snakes is more upturned than that of the eastern hognose. The threat and death-feigning displays of the southern hognose snake are less dramatic than those of the eastern hognose snake but are nonetheless characteristic of the species.

WHAT DO THE BABIES LOOK LIKE? The babies are miniatures of the adults.

OTHER NAMES Like the eastern hognose, this species is called spreading adder or puff adder in some regions.

DISTRIBUTION AND HABITAT Southern hognose snakes historically occurred in parts of the Coastal Plain of southern North Carolina, South Carolina, most of Florida, Georgia, and Alabama into southern Mississippi. No specimens have been documented from Alabama or Mississippi

Southern Hognose Snake

Heterodon simus

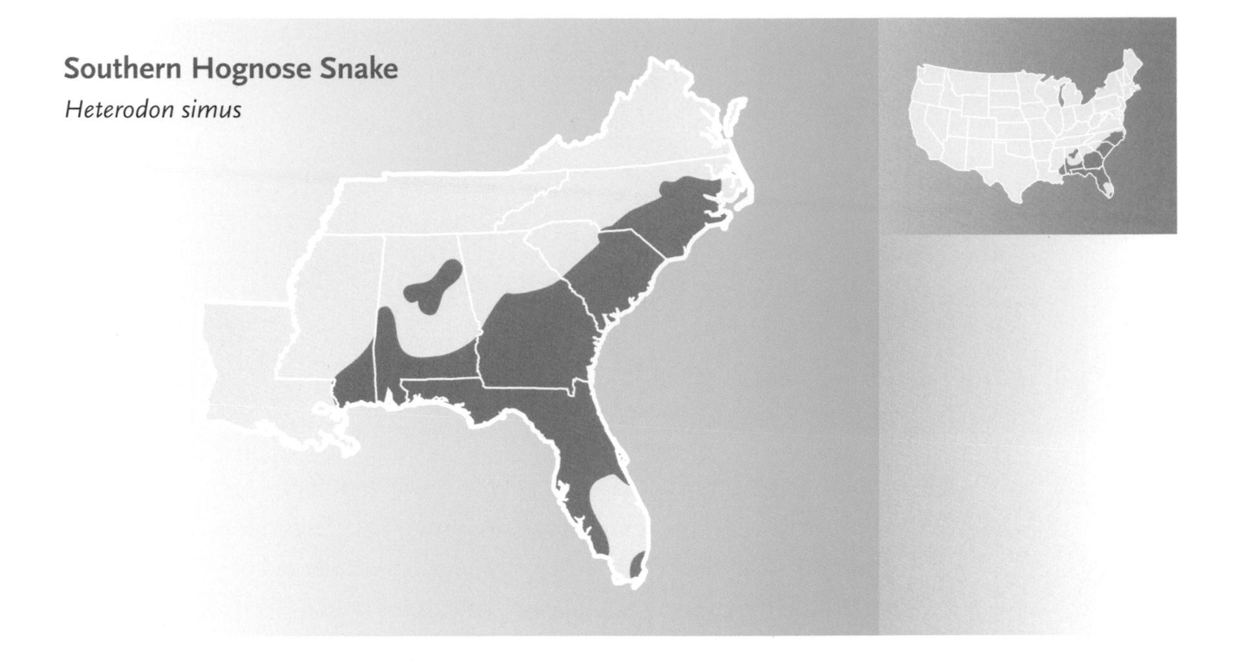

since the 1970s, and the species is assumed to be extirpated from those states. These snakes favor dry, well-drained, sandy soils and can be found in longleaf pine forests, abandoned agricultural fields, and a variety of scrubby oak woodlands, and are sometimes found in coastal sand dunes or even in cultivated fields.

BEHAVIOR AND ACTIVITY Southern hognose snakes may be found aboveground in every month, including sunny days in winter, and are active only during the daytime hours. Because of their strict association with sandy habitats and their upturned snouts, southern hognose snakes are assumed to be active burrowers that spend much of their time underground.

Occasional specimens of the southern hognose snake are reddish.

Southern hognose snakes are found almost exclusively in areas with sandy soil.

The diet of the southern hognose snake consists primarily of toads.

FOOD AND FEEDING Like their larger relative, southern hognose snakes eat primarily toads, including spadefoot toads. They use the upturned snout to dig up toads buried in loose soil and use the enlarged posterior teeth to puncture them. Southern hognose snakes sometimes feed on frogs, small lizards, and baby mice. They may subdue prey with their toxic saliva.

REPRODUCTION Not much is known regarding courtship, mating, or other aspects of the reproductive biology of southern hognose snakes. They apparently mate in late spring and early summer. Six to 14 (average = about 9) white, elongated eggs are laid in mid-to-late summer, and the young typically hatch in September or October.

PREDATORS AND DEFENSE Natural predators include various snakes, birds, and mammals associated with sandhill habitats. Nonnative fire ants, which also are found in sandy areas, may be a threat to the eggs and hatchlings. The threat response is similar to that of the eastern hognose but somewhat less dramatic, and includes spreading the neck and playing dead.

CONSERVATION The southern hognose snake has been extirpated or is in danger of extinction throughout much of its geographic range and was once proposed for federal listing under the Endangered Species Act. No specimens have been found in Alabama or Mississippi since the 1970s in areas where they once occurred, and they have disappeared from two-thirds of the counties where they were formerly found in the Carolinas, Georgia, and Florida. Nonnative fire ants have been implicated in their decline, but the steady loss and fragmentation of upland sandhill habitats through development is unquestionably a problem for the species in all areas. Like the eastern hognose snake, the southern hognose is frequently killed by vehicles when crossing highways in rural areas.

Mole kingsnake from Spartanburg, South Carolina

How do you identify a mole or prairie kingsnake?

SCALES
Smooth

ANAL PLATE
Single

BODY SHAPE
Moderately robust to robust

BODY PATTERN AND COLOR
Variable from solid brown to brownish gray with dark blotches

SIZE

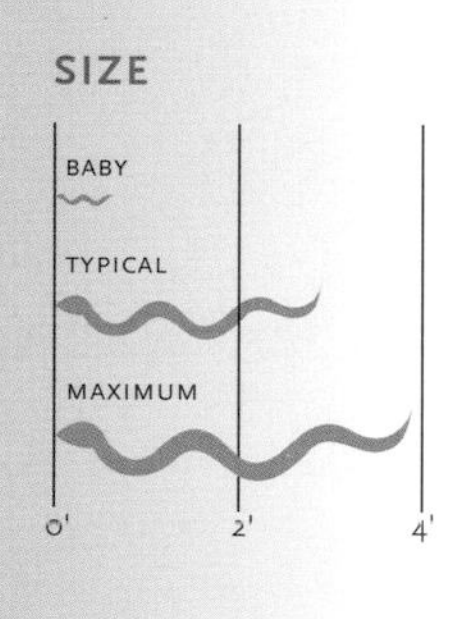

Mole and Prairie Kingsnakes

Lampropeltis calligaster

DESCRIPTION Mole and prairie kingsnakes have a rounded head, moderately robust body, and relatively short tail. All three subspecies (mole kingsnake, *L. c. rhombomaculata*; prairie kingsnake, *L. c. calligaster*; and South Florida mole kingsnake, *L. c. occipitolineata*) typically have a light brown body with reddish or brownish blotches down the back. The blotches of some mole kingsnakes become very faded with age, resulting in a nearly solid brown snake. The belly is lighter than the back and often has a faint but darker blotching pattern. The subspecies are distinguished by differences in numbers of blotches and scale counts. Intergradation occurs in the zone of contact between the mole and prairie kingsnakes (see map).

WHAT DO THE BABIES LOOK LIKE? Babies have distinctive blotches on the back.

DISTRIBUTION AND HABITAT Prairie or mole kingsnakes occur in each of the southeastern states, although they are sporadically distributed in Florida and absent from most of the southern portions of Georgia, Alabama, and Louisiana. They live in a variety of habitats but are most abundant in open

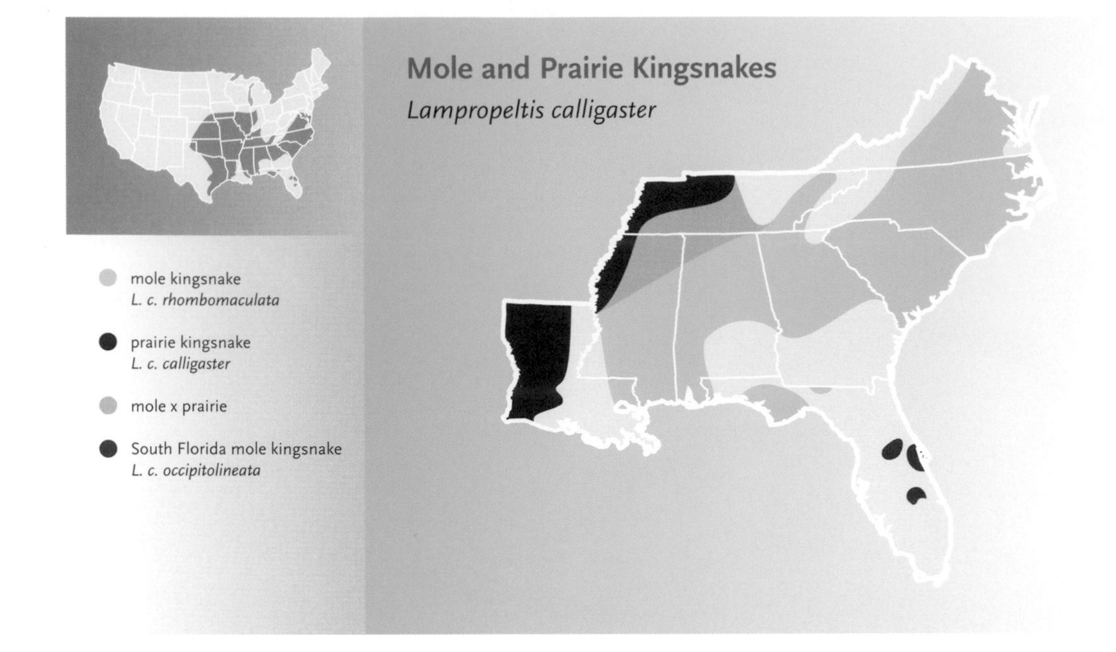

areas such as old agricultural fields, grasslands, savannas, and the edges of pastures. They can frequently be found beneath debris such as the old boards and discarded sheets of tin that are sometimes strewn around old barns and abandoned houses.

BEHAVIOR AND ACTIVITY Mole and prairie kingsnakes are typically active from spring to fall in cooler regions. They hibernate through the winter in most of the range, but may be active on warm days in the southern part of the range. They presumably spend much of the time underground in rodent burrows but are active aboveground during the day in the cooler parts of the year; during the summer months they are more likely to be found in early evening or at night.

The prairie kingsnake is found in western areas of Louisiana, Mississippi, and Tennessee.

FOOD AND FEEDING Prairie and mole kingsnakes eat shrews, mice, moles, pocket gophers, voles, racerunner lizards, skinks, and fence lizards, which they kill by constriction. They sometimes eat frogs, other snakes, and bird eggs. Though less well known than their larger relative, the common kingsnake, they have been reported to eat small pit vipers and are presumably immune to the effects of pit viper venom.

REPRODUCTION The breeding season begins in spring and lasts through the month of June. During mating, the male may bite the female on the neck. As many as 20 eggs (typically 8–14) are laid during summer. Mole kingsnakes usually lay smaller eggs than do prairie kingsnakes. The eggs hatch after an incubation period of about 2 months.

PREDATORS AND DEFENSE Common kingsnakes, large hawks, and carnivorous mammals are predators of mole and prairie kingsnakes, which, when captured may vibrate the tail, release musk, and bite. They settle down quickly in captivity and ordinarily are docile when handled.

Mole kingsnakes may be active aboveground during the day in cooler parts of the year.

CONSERVATION The species is uncommon throughout much of the Southeast, and in some cases it is difficult to know whether a population has been eliminated from a particular area or was never present. These kingsnakes are frequently killed on highways in agricultural areas and are probably victims of soil tillers and similar ground-breaking agricultural equipment.

The South Florida mole kingsnake occurs in scattered localities in central Florida.

Like the scarlet kingsnakes the Louisiana milksnakes have rings that encircle the body.

How do you identify a scarlet kingsnake or milksnake?

SCALES
Smooth

ANAL PLATE
Single

BODY SHAPE
Slender to moderately robust

BODY PATTERN AND COLOR
Scarlet kingsnake and Louisiana milksnake: ringed completely around the body with red, yellow, and black; milksnakes: light colored with darker reddish or brown blotches down the back and on the sides

DISTINCTIVE CHARACTERS
Louisiana milksnakes have a black nose; scarlet kingsnakes have a red nose

SIZE

BABY

TYPICAL

MAXIMUM

0' 2' 4'

SCARLET KINGSNAKE & LA. MILKSNAKE

EASTERN & RED MILKSNAKE

Scarlet Kingsnake and Milksnake

Lampropeltis triangulum

DESCRIPTION Milksnakes and scarlet kingsnakes are relatively robust, although smaller males may be less robust than females, which sometimes are slender as well. Some of the subspecies in the Southeast vary more dramatically from each other in appearance than some distinctly different snake species vary from each other. The scarlet kingsnake (*L. t. elapsoides*) and Louisiana milksnake (*L. t. amaura*) characteristically have brightly colored rings of red, yellow, and black completely encircling the body. The eastern milksnake (*L. t. triangulum*) is light colored, ranging from gray to light brown, with dark red or reddish brown blotches down the back and on the sides. The red milksnake (*L. t. syspila*) has a gray or whitish body with large red blotches down the back. A mixing of color patterns occurs in zones of intergradation between the scarlet kingsnake and eastern milksnake and among all three where they both come in contact with the red milksnake (see map).

WHAT DO THE BABIES LOOK LIKE? The young of all look like the adults, except that the red blotches on baby eastern milksnakes are more brightly colored than those on adults.

DISTRIBUTION AND HABITAT Milksnakes and scarlet kingsnakes are found throughout every southeastern state in a variety of habitats. Louisiana milksnakes and scarlet kingsnakes are most abundant in forests, where they often hide beneath the loose bark of dead trees, especially pines. Eastern and red milksnakes favor more open habitats and can frequently be found under debris in abandoned agricultural areas, meadows, and along the edges of woodlands.

BEHAVIOR AND ACTIVITY This species is active in all warm months throughout the Southeast and in the winter months during warm periods. Individuals can be found on the surface during both night and day but are decidedly more active at night in warm regions. The scarlet kingsnake is noted for hiding beneath the loose bark of dead pine trees, especially during early spring and autumn.

FOOD AND FEEDING Milksnakes and scarlet kingsnakes feed mostly on lizards, small rodents, other snakes, and eggs. Prey may be detected by smell and actively hunted or ambushed, and is killed by constriction. Milksnakes, being larger, eat larger prey than scarlet kingsnakes, which feed primarily on lizards.

Did you know?

In many species of snakes, females get much larger than males.

Scarlet kingsnakes are powerful constrictors for their size.

Scarlet kingsnake from the South Carolina Coastal Plain

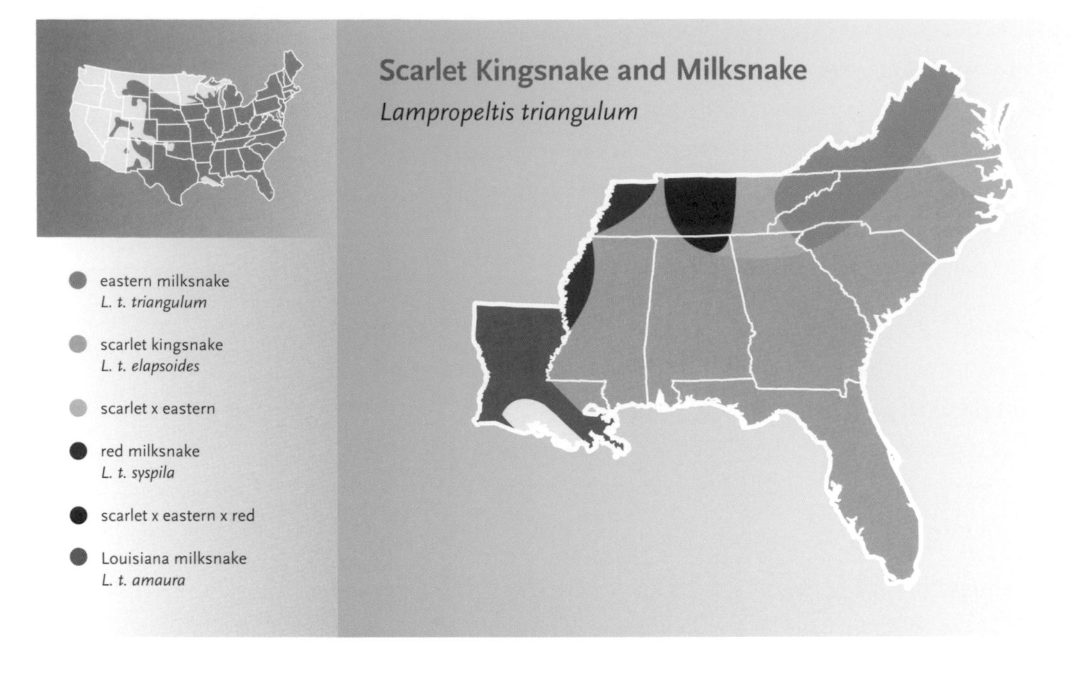

Scarlet kingsnake emerging from egg

REPRODUCTION The mating season lasts from early April through May. Some individuals also mate in the fall, and the female stores the sperm until fertilization occurs the following spring. Females leave pheromone trails that males follow, and during courtship the male bites the neck of the female and rhythmically moves his body against hers. Eggs are laid during early summer inside rotting logs, under moist bark, and in piles of rotting vegetation. Eastern milksnakes lay clutches of 1–24 eggs (average = about 9); the smaller races (Louisiana milksnake and scarlet kingsnake) usually lay only 4 or 5 eggs. The eggs hatch in late summer or early fall.

Red milksnake from Tennessee

Eastern milksnakes have blotches instead of rings.

PREDATORS AND DEFENSE The primary predators of milksnakes and scarlet kingsnakes are snake-eating snakes, birds of prey, and carnivorous mammals. Scarlet kingsnakes and Louisiana milksnakes resemble the eastern coral snake with their red, yellow, and black rings, and some herpetologists consider this resemblance to be a form of protective mimicry that makes some predators, especially birds, hesitate before attacking. Milksnakes and scarlet kingsnakes usually vibrate the tail, release musk, and bite and chew when captured.

Unlike coral snakes, scarlet kingsnakes and milksnakes have black bands that separate red and yellow bands.

CONSERVATION The conversion of woodland habitats to urban areas is harming this forest-dwelling species throughout its geographic range. At one time, scarlet kingsnakes were collected in large numbers for the pet trade, and according to some sources this practice continues in some parts of Florida. However, the wide variety of captive-bred kingsnakes now available to pet owners has considerably reduced field collecting.

LARGE TERRESTRIAL SNAKES

Eastern kingsnake from Gaston County, South Carolina

Common Kingsnake *Lampropeltis getula*

DESCRIPTION The back is typically black or dark brown with yellow or white markings (bands or spots); the belly is usually shiny black or mottled with yellow or white. The commonly recognized kingsnake subspecies in the Southeast are the eastern kingsnake (*L. g. getula*), with thin white or yellow bands across the back; the speckled kingsnake (*L. g. holbrooki*), with a yellow spot on most or all of the black body scales; and the eastern black kingsnake (*L. g. niger*), with limited spotting on the sides and banding absent or noticeable only as faint spotting. The Florida kingsnake (*L. g. floridana*) has scales that are predominantly yellowish brown or cream colored with a faint banding pattern visible along the body; the belly is light colored with slightly darker spots. Eastern kingsnakes and Florida kingsnakes interbreed in some regions of central Florida to produce specimens that have variable color patterns.

Some herpetologists recognize the Outer Banks (North Carolina) kingsnake (*L. g. sticticeps*), with its more brownish appearance and yellow spotting between the bands, and

Some herpetologists consider the Apalachicola kingsnake to be a distinct subspecies.

How do you identify a common kingsnake?

SCALES
Smooth

ANAL PLATE
Single

BODY SHAPE
Robust; rounded head; relatively short tail

BODY PATTERN AND COLOR
Shiny black with white or yellow speckles or rings

DISTINCTIVE CHARACTERS
Scales usually shiny

SIZE

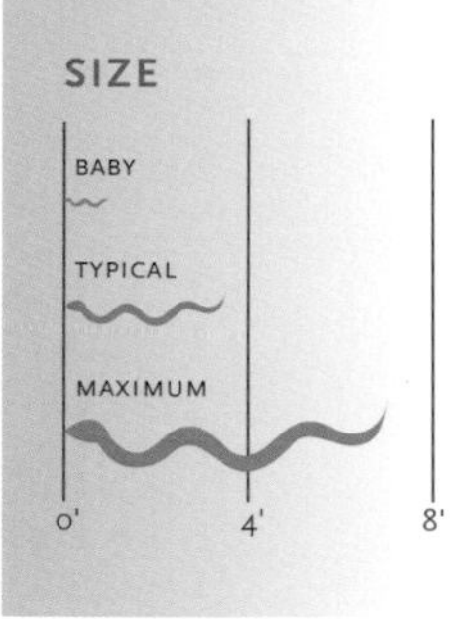

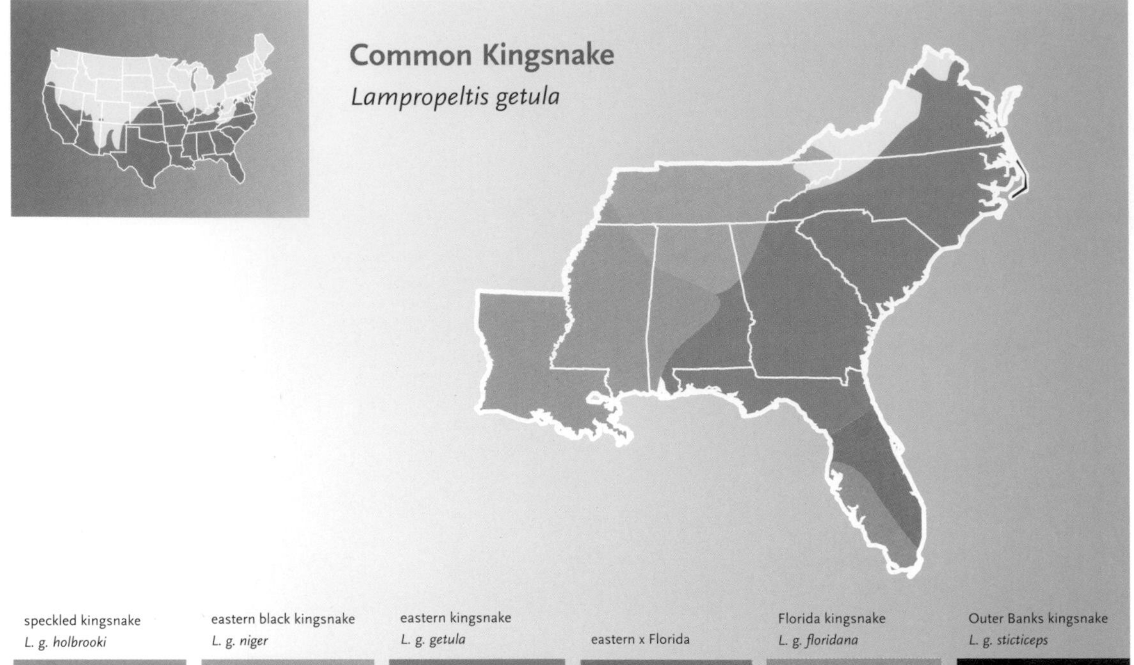

Common Kingsnake

Lampropeltis getula

speckled kingsnake
L. g. holbrooki

eastern black kingsnake
L. g. niger

eastern kingsnake
L. g. getula

eastern x Florida

Florida kingsnake
L. g. floridana

Outer Banks kingsnake
L. g. sticticeps

Did you know?

Common kingsnakes are immune to the venom of pit vipers such as copperheads, cottonmouths, and rattlesnakes. Whether kingsnakes are immune to the venom of coral snakes is still unknown.

the profusely yellow-spotted and sometimes banded Apalachicola kingsnake (*L. g. goini*) as distinct subspecies. Intergrades occur in areas of contact between eastern and Florida kingsnakes (see map).

WHAT DO THE BABIES LOOK LIKE? Baby common kingsnakes generally look like miniature versions of the adults, although hatchlings of all the southeastern subspecies may have visible yellow or white bands and minimal spotting.

OTHER NAMES Colloquial names for the common kingsnake vary regionally because of the many color patterns the species exhibits, but chain kingsnake is a name frequently used for the eastern kingsnake.

DISTRIBUTION AND HABITAT Kingsnakes are found throughout all or most of every southeastern state in a number of habitat types, including hardwoods, sandhills, grassy meadows, pine forests, savannas, and flatwoods. Kingsnakes also persist in suburban and agricultural areas. Although characteristically terrestrial, they often inhabit grassy shorelines of wetlands, including swamps and streams.

Black kingsnakes characteristically have few or no yellow spots on the back.

BEHAVIOR AND ACTIVITY Kingsnakes are active primarily in the daytime, although early evening activity has been observed during summer. Nocturnal movement is relatively uncommon. Seasonal activity is similar to that of other southeastern reptiles in that they become inactive in late fall, hibernate through the winter, and become active again in early spring. Kingsnakes will bask in the sun alongside burrows or on low vegetation during cool weather and may be found in or beneath rotting logs or under other debris at all times of the year.

FOOD AND FEEDING Common kingsnakes are strong constrictors well known for eating other snakes, including venomous species. They are immune to the venom of pit vipers and thus readily consume rattlesnakes, copperheads, and cottonmouths in addition to any other snake their size or smaller. Common kingsnakes also eat rodents, lizards, reptile eggs, birds and their eggs, and even small turtles. When eating small or defenseless prey, they often forgo constriction and simply swallow the prey alive. Common kingsnakes find their prey primarily through smell, although vision may be important as well.

Did you know?

Kingsnakes make excellent pets for children or adults interested in keeping a snake in captivity. A variety of striking color patterns can be obtained through reputable reptile breeders in many areas.

Typical speckled kingsnakes have a yellow spot on each scale on the back.

REPRODUCTION Common kingsnakes mate in the spring. During courtship and mating the male may bite and hold on to the female's neck. Approximately 10 eggs (range = 3–24) are laid in June or July within rotting logs, stumps, or similar moist, protected spots. The eggs, which tend to stick together in a clump, hatch 2–2.5 months later. Females are known to produce clutches of eggs fertilized by multiple males.

PREDATORS AND DEFENSE Kingsnakes face the same array of natural predators as most other snakes, but because of their relatively large size, adults need fear only the largest predators. When threatened in the wild, kingsnakes may vibrate the tail, assume a striking position, and release musk, and often bite if picked up or harassed. However, these defensive behaviors usually soon disappear and they can be easily handled.

Although most make good pets, many common kingsnakes will defend themselves by biting when first captured.

CONSERVATION Common kingsnakes are not protected by the federal government. The Outer Banks kingsnake, recognized by some herpetologists as a separate subspecies, has been listed as a Species of Special Concern by the state of North Carolina. Because kingsnakes are easily recognized and have a reputation for eating venomous snakes, they are sometimes spared the fate of many snakes that encounter people, although many are killed by vehicles on roadways. Overcollecting for the pet trade has been suggested by some herpetologists as the cause of a possible decline of the Apalachicola kingsnake in western Florida.

Florida kingsnake from Hillsborough County, Florida

Northern pine snake from North Carolina

Pine Snake — *Pituophis melanoleucus*
Louisiana Pine Snake — *Pituophis ruthveni*

DESCRIPTION With the exception of the subspecies known as the black pine snake (*P. m. lodingi*), pine snakes characteristically have dark blotches on a lighter background. The other two subspecies are the northern pine snake (*P. m. melanoleucus*) and the Florida pine snake (*P. m. mugitus*), in which the dark markings may be less visible, especially on the front part of the body. The pattern varies among regions and among individuals within regions, being lighter and more muted in some areas and more darkly contrasted in others. Adult black pine snakes are black above and below. The Louisiana pine snake (*P. ruthveni*) resembles its eastern counterpart.

WHAT DO THE BABIES LOOK LIKE? Baby pine snakes generally look like the adults, but black pine snake babies are lighter colored than the adults and show signs of dark blotches instead of being solid black. Newborn pine snakes are longer and stouter than many southeastern snakes are as adults.

DISTRIBUTION AND HABITAT Pine snakes are found in all of the southeastern states but are patchily distributed over much of their range. The two eastern subspecies collectively range fairly continuously throughout South Carolina, southeastern Georgia, and all but the southern portion of Florida.

How do you identify a pine snake?

SCALES
Keeled

ANAL PLATE
Single

BODY SHAPE
Heavy bodied, narrow, pointed head

BODY PATTERN AND COLOR
Light colored above with dark blotches; white belly

DISTINCTIVE CHARACTERS
Large scale covering a pointed nose

SIZE

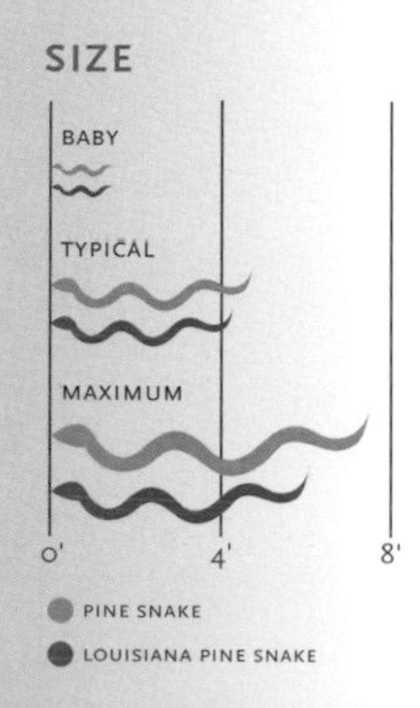

Black pine snake from southern Alabama

Louisiana pine snake

Only scattered localities are known for the species in Virginia, North Carolina, and Tennessee. The black pine snake ranges from southern Alabama and Mississippi to Louisiana. The Louisiana pine snake is found in west-central Louisiana and into east-central Texas and is geographically isolated from its relatives the eastern pine snakes and the western bullsnakes. Pine snakes generally live in areas with well-drained, sandy soil where they can easily burrow in search of prey. They can be found in longleaf pine sandhill areas, pine barrens, scrub oak, dry rocky areas in the mountains, and abandoned agricultural fields. They spend much of their time underground, and they usually do not climb.

BEHAVIOR AND ACTIVITY Pine snakes are inactive from late fall to spring throughout their geographic range, and hibernate in most regions. During the active season they spend most of their time underground in burrows that they dig themselves or were dug by small mammals or gopher tortoises, or in tunnels left by decayed roots. They are most active on the surface in the spring but may be found aboveground during the warm months in many parts of the Southeast. Surface activity occurs almost exclusively during daylight hours. Pine snakes may have very large home ranges—up to several hundred acres—and individuals sometimes move long distances overland.

FOOD AND FEEDING Pine snakes are extraordinarily powerful constrictors that consume rabbits, squirrels, rats, mice, and pocket gophers. They forage actively both above and below the ground. Small prey items (e.g., juvenile rodents) are eaten alive, without constriction. They will occasionally eat bird eggs.

REPRODUCTION Pine snakes mate in the spring, and males presumably find females by following their pheromone trails. When a male finds a female, he will rub alongside her and may bite her neck to prevent escape. During summer, females dig long burrows in loose soil and lay eggs at

Florida pine snakes are not as vividly marked as the other subspecies.

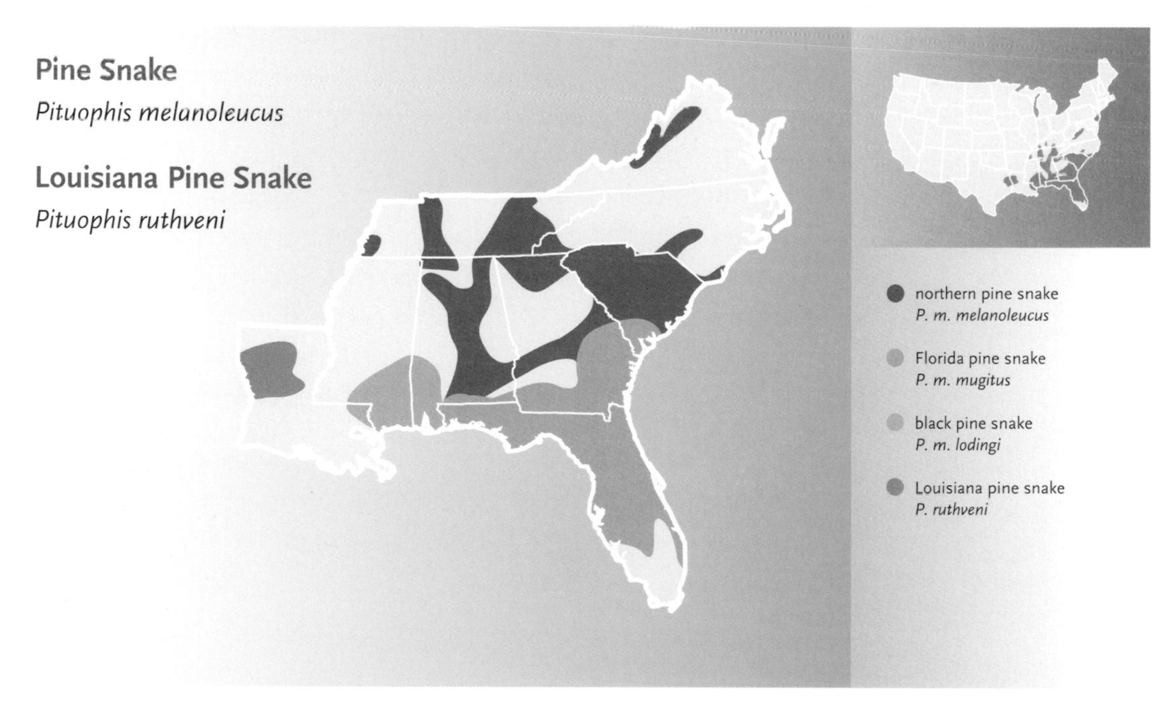

Pine snakes often hiss loudly when threatened.

the end. Egg clutches range in size from 3 to 24 eggs (usually = 9–10). Females nest communally in some portions of the geographic range, so that several clutches of eggs may be found in the same burrow. The unusually large babies hatch about 2.5 months later.

PREDATORS AND DEFENSE Predators include large birds of prey and larger mammals such as coyotes. Pine snakes encountered in the wild often hiss loudly, sometimes from several feet away, making a sound that is similar to that of a large rattlesnake. When picked up by a person, they may coil tightly around their captor's arm. Occasionally pine snakes will bite.

Did you know?

Pine snakes make a hissing sound that resembles the whirring sound of a large, rattling rattlesnake.

CONSERVATION Habitat destruction for commercial development is probably the single greatest threat to pine snakes. Because of their large size, generally slow crawling speed, and tendency to be active during peak daytime traffic periods, pine snakes are particularly vulnerable to road mortality. Hence, rural roads are among the greatest hazards faced by pine snakes in many parts of their geographic range. Of more than 100 pine snakes recorded in a South Carolina study, nearly 85% were snakes observed crossing or already dead on highways. Pine snakes are also victims of intentional human persecution because of their superficial resemblance to rattlesnakes. The Louisiana pine snake and black pine snake have both been proposed as candidate species for listing under the Endangered Species Act.

SCIENTIFIC NOMENCLATURE Herpetologists disagree about the taxonomy of the Louisiana pine snake; some consider it a separate species, and others view it as a subspecies of *P. melanoleucus* or of the bullsnake (*P. catenifer*).

Black rat snake from Davidson, North Carolina

Rat Snake

Elaphe obsoleta

DESCRIPTION Rat snakes range from slender to relatively stout; some seem relatively slim, like racers, and others are heavier bodied. Presumably the size reflects the individual's recent diet. Few North American snakes show as much regional variation in body pattern and color as the rat snakes do, from the almost solid-colored black rat snake (*E. o. obsoleta*) found in northern parts of the Southeast to the yellow rat snake (*E. o. quadrivittata*), which is yellow with black stripes in Florida and more greenish with black or brown stripes in coastal areas of Georgia and the Carolinas. The Everglades rat snake (*E. o. rossalleni*) ranges from bright orange to yellowish orange with indistinct striping. The widespread gray rat snake (*E. o. spiloides*) characteristically has a very light to dark gray body with large, darker gray blotches down the back and smaller ones along the sides. The subspe-

A rat snake from Aiken, South Carolina.

How do you identify a rat snake?

SCALES
Weakly keeled

ANAL PLATE
Usually divided

BODY SHAPE
Slender to moderately stout and shaped in cross section like a loaf of bread

BODY PATTERN AND COLOR
Solid black, or with blotches or stripes

DISTINCTIVE CHARACTERS
Shaped in cross section like a loaf of bread

SIZE

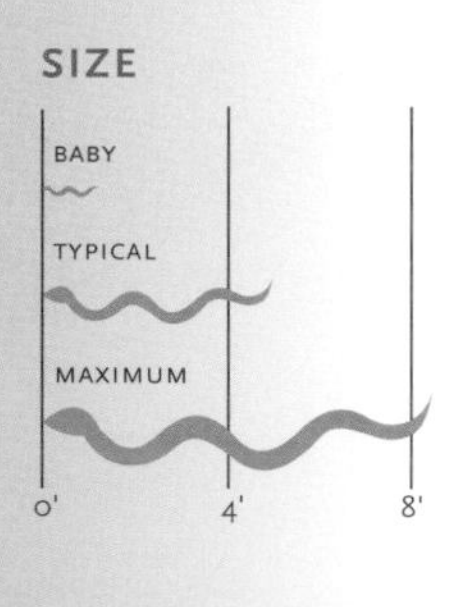

cies of rat snake prevalent throughout most of southern Louisiana, the Texas rat snake (*E. o. lindheimerii*), resembles the gray rat snake in pattern, but some specimens are more brown. The belly is generally similar in color to the back, although black rat snakes often have white under the front part of the body. Several zones of intergradation are apparent where black, Texas, gray, and yellow rat snakes have contiguous ranges with one of the other subspecies (see map).

WHAT DO THE BABIES LOOK LIKE? Newborn rat snakes from most areas look like miniature versions of the gray or Texas rat snake, with dark gray or brown blotches on a lighter gray body. They assume their regional adult coloration and pattern as they grow older. Baby Everglades rat snakes may be yellowish with less distinctive blotches.

Baby black rat snakes are patterned like gray rat snakes.

OTHER NAMES Rat snakes are traditionally called chicken snakes throughout most of the South, but oak snake and goose snake are commonly used names in some localities. Along with the superficially similar-looking black racer, the black rat snake is usually just called the black snake, but is sometimes referred to as the pilot black snake in reference to a popular belief that they lead rattlesnakes to their mountain dens for the winter.

DISTRIBUTION AND HABITAT Rat snakes are native throughout every southeastern state and live in a wide variety of wooded and forested habitats. They are common in many suburbs, especially if the area still has many large trees in which they can climb. They also frequent abandoned buildings and barns where they can hunt rodents and birds. They are often found near the edges of forests, and are sometimes found in swamp forests and other wetlands.

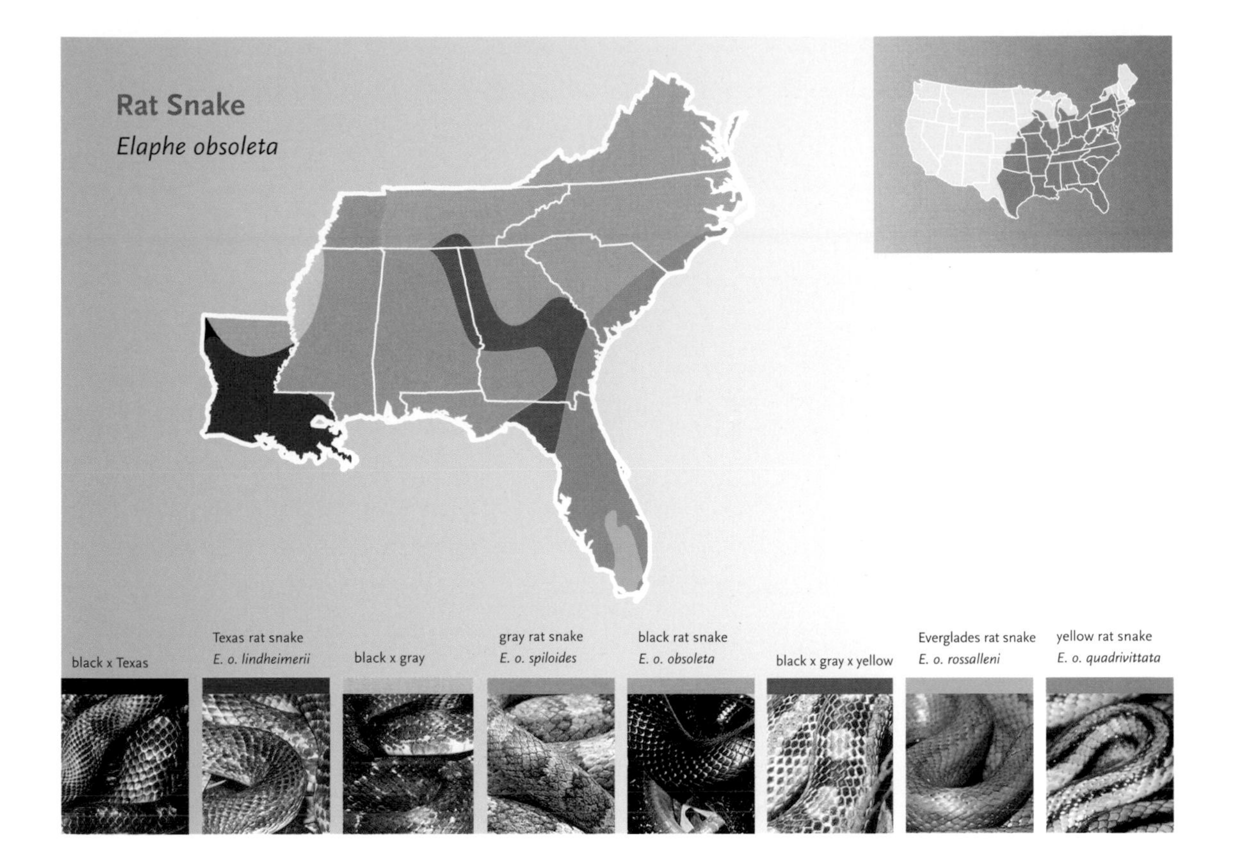

BEHAVIOR AND ACTIVITY Rat snakes can be found in the Southeast during most times of the year, but during extremely cold or prolonged winters they may hibernate for 2–4 months in rotten stumps, hollow trees, or abandoned houses. During spring they are often active on the ground, in trees, and in flooded swamps. They are generally most active in daytime, especially when climbing in trees and bushes in the spring, but may be active at night during the summer months.

FOOD AND FEEDING Rat snakes are constrictors and active hunters that use both vision and smell to find their prey. Adults prey primarily on small mammals, birds, and bird eggs, which they either swallow whole or break inside the

Everglades rat snake from Monroe County, Florida

Greenish version of yellow rat snake from coastal South Carolina

throat by squeezing the egg against their vertebrae; the broken shells are either regurgitated or swallowed. Young rat snakes feed mostly on treefrogs, small lizards, and baby rodents.

REPRODUCTION Rat snakes are egg-layers that generally mate during April, May, and June, and possibly in the fall in some areas. When one male encounters another during the breeding season, they sometimes engage in a wrestling match, presumably in competition for a female. During courtship and mating, the male frequently bites the female's neck. The clutch size ranges from 4 to 44 eggs, but the usual number is about 15. Eggs are laid in stump holes, tree holes, or other dark, moist situations, and several females may nest together. Rat snakes have been known to return to the same area to nest year after year. The eggs hatch in about 2 months.

Yellow rat snake from Alachua County, Florida

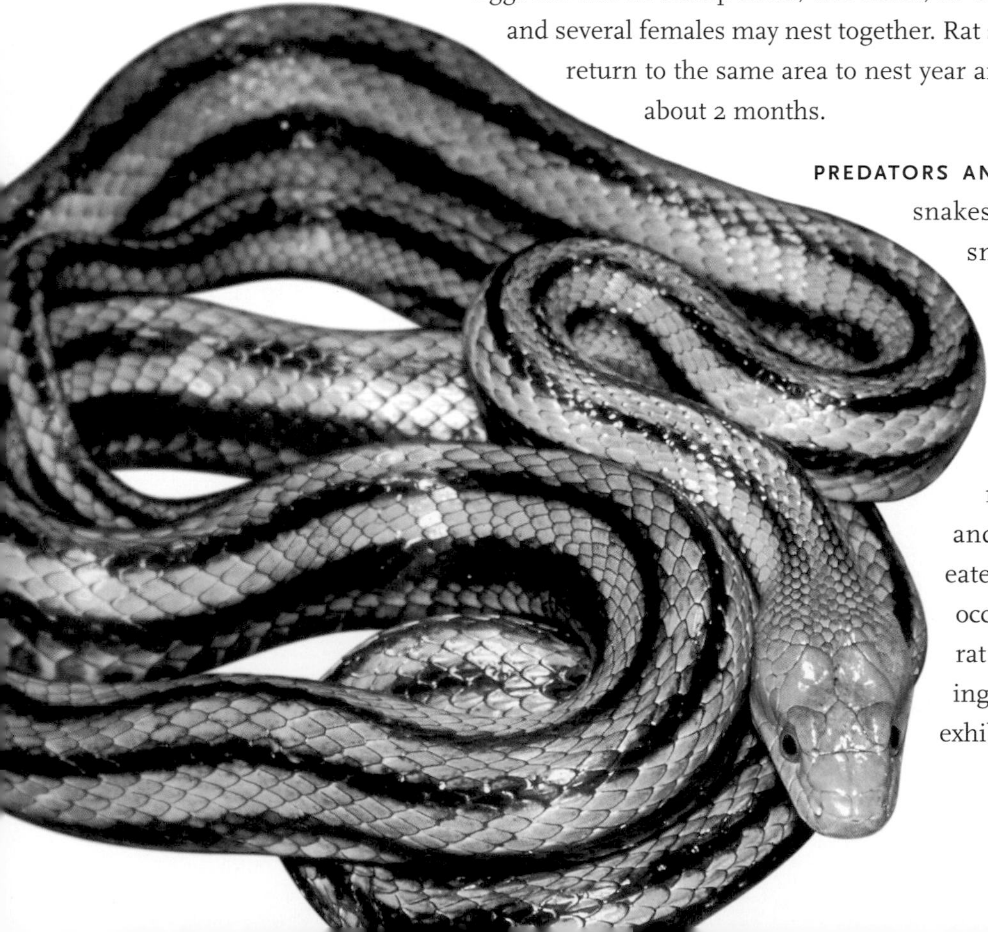

PREDATORS AND DEFENSE A variety of snakes (e.g., copperheads, kingsnakes, racers, and indigo snakes), birds of prey (both hawks and owls), and mammals (raccoons, weasels, and coyotes) have been reported to prey on rat snakes. Rat snake eggs and juveniles are commonly eaten by native predators that occupy the same habitats. Wild rat snakes encountered crawling on the ground commonly exhibit a behavior called "kink-

Gray rat snake from Georgia

This South Carolina rat snake shows characters intermediate between those of yellow and black rat snakes.

ing," in which the stretched-out snake makes a series of irregular kinks along the length of its body and then remains motionless; presumably this configuration serves as a form of camouflage. Rat snakes may bite when picked up, but if handled gently, many do not. They also may wrap around the captor's arm or the head of a predator and may exude a very unpleasant musk from their cloacal (anal) glands.

CONSERVATION Small rat snakes in suburban areas are killed by domestic cats and dogs. Many adult rat snakes are victims of human persecution. Rat snakes are frequent inhabitants of both suburban and farm communities, and will not hesitate to eat chicken and duck eggs or small chickens and ducklings. Rat snakes caught in the act are often killed. Rat snakes climbing trees in pursuit of bird eggs are likely to attract the attention of birds, whose mobbing cries are likely to be noticed by people. Along with bird eggs, however, rat snakes eat enormous numbers of rats and mice wherever they live, and people who cannot appreciate their beauty or respect their right to share the land should value them as rodent-control agents. One of the major threats to rat snakes—an irrational fear by some people—can be lessened by educating the public to understand that recognizing another species as a predator is insufficient justification for a person to kill it.

The range of Texas rat snakes extends eastward into Louisiana.

SCIENTIFIC NOMENCLATURE The taxonomy of rat snakes has been under intensive review by herpetologists. Some herpetologists place the New World rat snakes in the genus *Pantherophis* rather than *Elaphe*. Additionally, genetic studies indicate that the current species *E. obsoleta* may actually be composed of three distinct species: the eastern rat snake (*E. alleghaniensis*), the Texas rat snake (*E. obsoleta*), and the gray rat snake (*E. spiloides*), with no recognized subspecies.

Corn snake from the Coastal Plain of South Carolina

Corn Snake

Elaphe guttata

DESCRIPTION Some corn snakes are among the most colorful snakes in the eastern United States, being bright red or orange with darker blotches down the back and sides, and others are duller variations of orange and brown, but all have two black-bordered stripes that extend forward across the head and connect to form a point between the eyes. A stripe behind the eye extends onto the scales of the neck. In western Louisiana, the corn snake may *hybridize* with the Great Plains rat snake (*Elaphe emoryi*), resulting in specimens that are more drab than the eastern forms. Coastal corn snakes are often more reddish than inland forms and are prominently marked with black.

Lines on the head of baby corn snakes form a point between the eyes.

WHAT DO THE BABIES LOOK LIKE? Baby corn snakes have bolder patterns than the adults, but usually less orange or red.

OTHER NAMES The corn snake is known as the red rat snake in some localities.

How do you identify a corn snake?

SCALES
Weakly keeled

ANAL PLATE
Divided

BODY SHAPE
Moderately proportioned, with a flat belly and shaped in cross section like a loaf of bread

BODY PATTERN AND COLOR
Body reddish, yellowish, orange, or brown, with large blotches on the back and smaller ones on the sides; belly a black-and-white checkerboard with varying amounts of red; and black stripes on the underside of the tail

DISTINCTIVE CHARACTERS
Lines on head form a point between eyes

SIZE

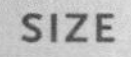

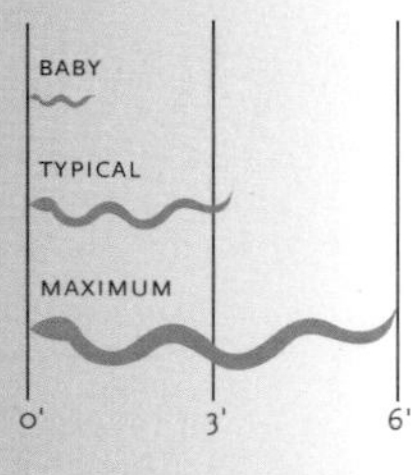

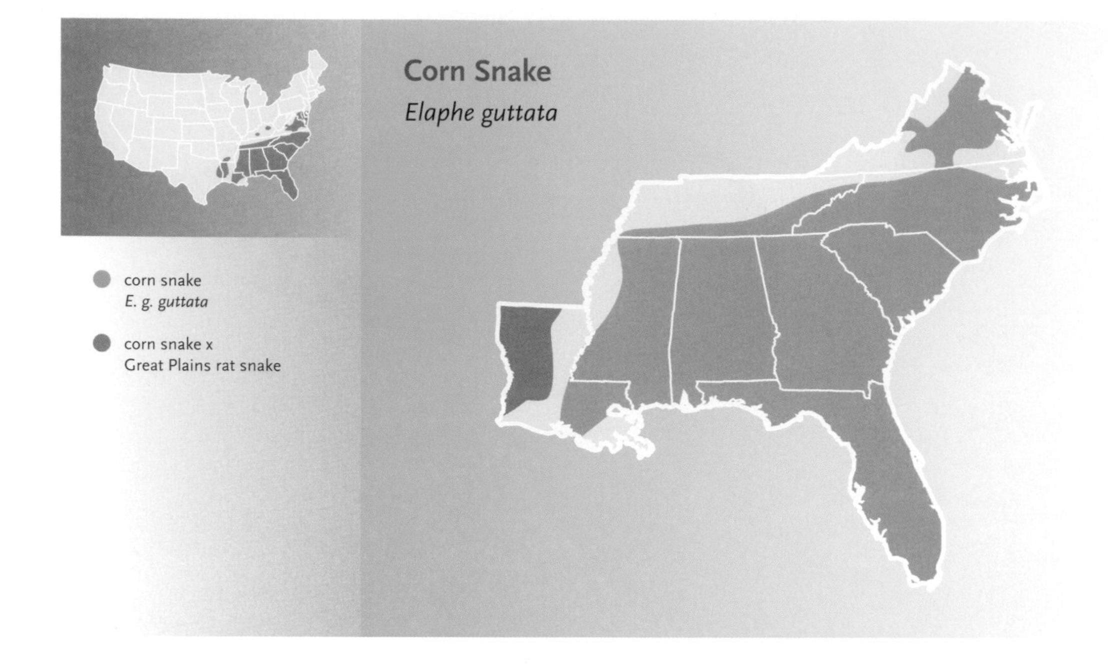

DISTRIBUTION AND HABITAT Corn snakes occur throughout all or a large part of every southeastern state in a variety of habitats, including pine stands, abandoned agricultural fields, and hardwood forests. They are sometimes found near human habitations and can be common under debris left around vacant houses and barns.

Many corn snakes have a checkerboard or "piano-key" belly pattern.

BEHAVIOR AND ACTIVITY Corn snakes hibernate for several weeks or months in the northern portions of the Southeast but are active virtually year-round in the southern and warm coastal regions. When first becoming active in the spring they confine their aboveground movements to the daylight hours, but they become active primarily at night during the summer. In the fall they are again more likely to be active during daylight hours. Corn snakes frequently climb and are often found beneath the bark of dead trees.

FOOD AND FEEDING The diet consists mostly of small mammals, but birds and their eggs are also eaten, as are lizards, other snakes, and frogs. Juveniles eat mostly small treefrogs and lizards. Corn snakes kill large prey by constriction and eat smaller animals alive. They are active foragers and may hunt underground for rodents, moles, and voles, which they locate primarily by odor, but they may also detect small mammals and birds by their body heat.

REPRODUCTION Like other members of the genus *Elaphe*, corn snakes are egg-layers. Mating begins in the early spring. While courting, the male

moves slowly along the female's back, stimulating her with undulations of his body. Eggs are laid in June and July over most of the range, but a little earlier in Florida. Clutches may number 3–40 eggs, but the typical clutch size is about 14. Larger females produce more eggs than smaller females. Hatching takes place about 2 months after the eggs are laid.

Corn snake from Leon County, Florida

PREDATORS AND DEFENSE Natural predators include larger mammals, birds of prey (both hawks and owls), and kingsnakes. When threatened, corn snakes generally try to escape, but if unable to do so may bite and vibrate the tail. Most herpetologists do

Corn snakes kill their prey by constriction.

Corn snakes in Louisiana, which interbreed with the Great Plains rat snake, are more drab in color than their eastern counterparts.

Corn snake from Spartanburg, South Carolina

not consider their musk to be as disagreeable as that of their close relative, the rat snake. Corn snakes occasionally bite when first picked up but usually settle down quickly in captivity.

CONSERVATION Aside from natural predators, one of the greatest threats to corn snakes is highway mortality. Occasionally corn snakes are killed unnecessarily by people who mistake them for copperheads.

SCIENTIFIC NOMENCLATURE Some herpetologists refer to the New World genus *Elaphe* as *Pantherophis*, and some consider corn snakes in western Louisiana to be a separate species, Slowinski's corn snake (*E. slowinskii*), on the basis of genetic studies.

This southern black racer from South Carolina exhibits the coloration typical of southern and northern black racers.

Racer

Coluber constrictor

DESCRIPTION Several subspecies of racers are found in the Southeast (five in Louisiana alone). Most are solid colored above and below. Three subspecies are totally black above and below: the northern black racer (*C. c. constrictor*) and southern black racer (*C. c. priapus*) have white chins, and the brown-chinned racer (*C. c. helvigularis*) has tan to brown lips. The Everglades racer (*C. c. paludicola*), blackmask racer (*C. c. latrunculus*), and eastern yellowbelly racer (*C. c. flaviventris*) are various shades of gray, green, or brown above and vary from light yellowish to bluish below. The buttermilk racer (*C. c. anthicus*) and closely related tan racer (*C. c. etheridgei*) differ from the solid-colored racers in having light spotting on a darker body. All racers are slender and streamlined for speed.

WHAT DO THE BABIES LOOK LIKE? Baby racers differ from the solid-colored adults in having light-colored bodies with dark blotches along the back and sides.

Baby racers have blotches and will assume the solid color characteristic of most subspecies as they mature.

How do you identify a racer?

SCALES
Smooth

ANAL PLATE
Divided

BODY SHAPE
Slender and streamlined

BODY PATTERN AND COLOR
Typically solid black, gray, or brownish; some with light spots

SIZE

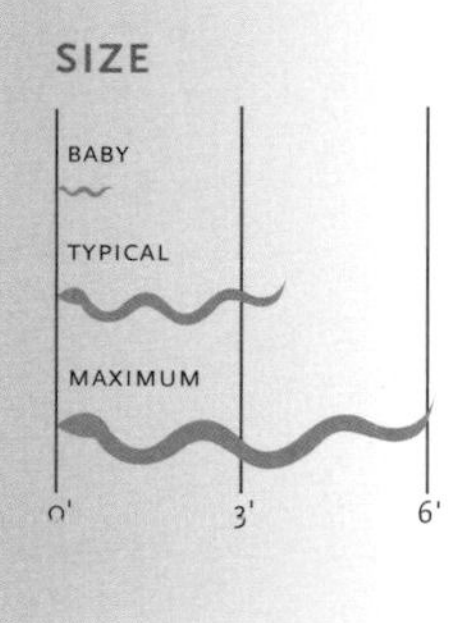

Black racers are lighter below than above.

OTHER NAMES Racers are often called black runners or blue runners in the rural South.

DISTRIBUTION AND HABITAT Racers occur throughout every state in the Southeast in a variety of habitats, but they prefer open areas such as old agricultural fields, forests with sparse undergrowth, brushy areas, and the edges of marshes and swamps. They often stay close to a retreat such as an animal burrow into which they can quickly retreat if disturbed.

Did you know?

Many nonvenomous snakes such as racers, kingsnakes, and rat snakes vibrate their tails when frightened. If the tail is vibrated in dry leaves, it may produce a sound much like a rattlesnake's rattle.

BEHAVIOR AND ACTIVITY Racers hibernate for at least a few weeks each winter in most of the Southeast north of Florida, and become active again in the early spring. They seek refuge during the winter, as well as at other times of the year, beneath leaves and other ground litter, under logs, in stumps, or in burrows and tunnels in the soil. Racers are active exclusively during the daytime, and they travel much longer distances than most other southeastern snakes do. Racers frequently climb into trees or bushes.

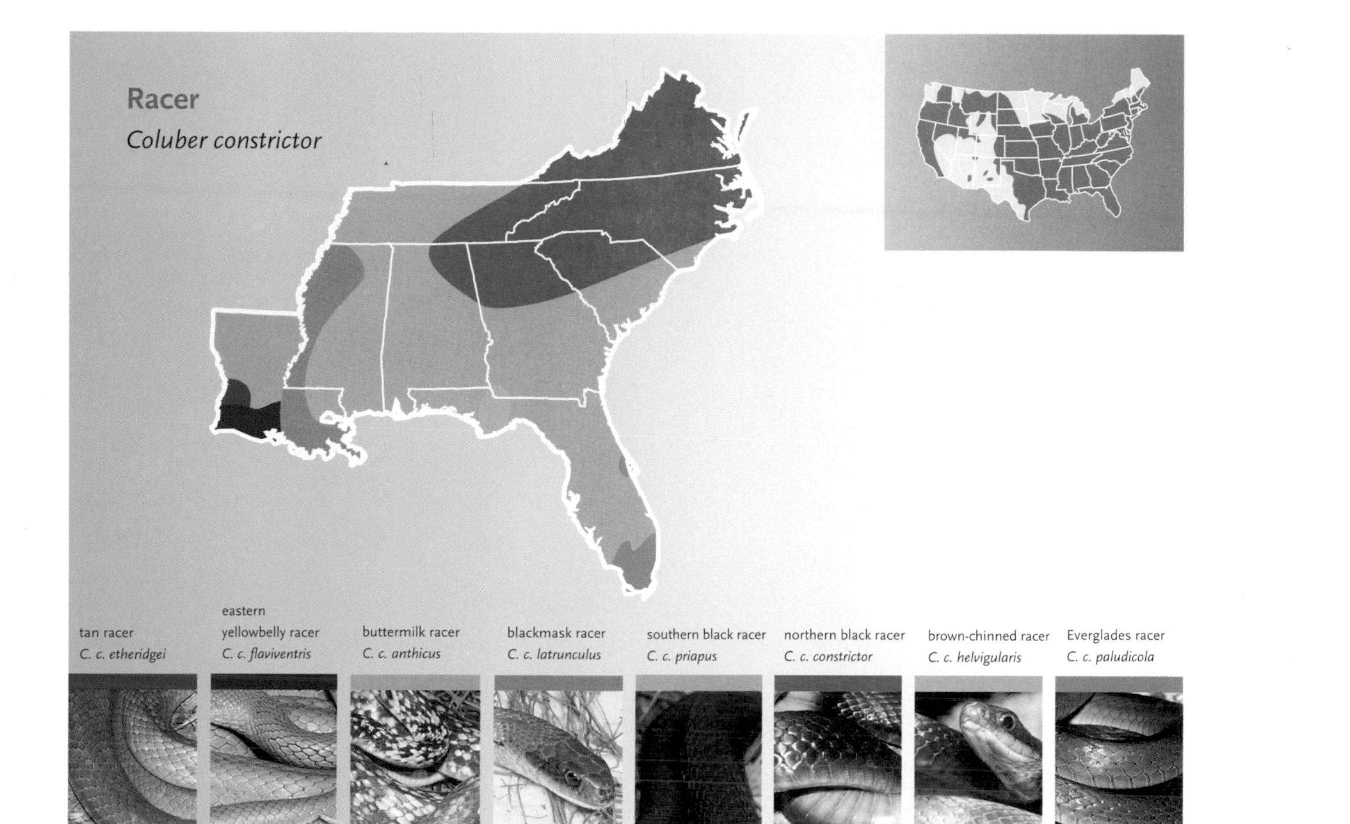

FOOD AND FEEDING Racers probably have a more diverse diet than any other North American snake. The wide variety of prey items includes small rodents, lizards, other snakes (including smaller racers), birds, frogs, insects, and even small turtles. Racers will eat the eggs of other reptiles and birds, and will even eat some venomous snakes such as copperheads and pigmy rattlesnakes. They are visually oriented, active hunters that often search for prey with the head held high like a periscope. When prey is spotted, they chase it down, seize it, and swallow it alive. Despite their scientific name, racers are not constrictors. Large or potentially dangerous prey may be chewed on until subdued and easier to swallow.

Everglades racer from Dade County, Florida

Buttermilk racers occur throughout much of Louisiana.

Racer eggs are covered by small granules that resemble salt.

REPRODUCTION Racers mate between April and July and lay 4–36 eggs (usually 10–13) during the summer. Larger females usually lay more eggs than smaller ones. The oblong eggs, which are covered by tiny granules that look like salt, are laid in moist, protected areas such as under fallen bark or large rocks, or buried a few inches underground. They hatch in about 2–2.5 months.

The geographic range of the eastern yellowbelly racer extends into southwestern Louisiana from Texas.

A blackmask racer from Terrebonne Parish, Louisiana

The brown-chinned racer is restricted to a small area of the panhandle of Florida and southern Georgia.

PREDATORS AND DEFENSE Natural predators include birds of prey, such as hawks, and larger mammals, such as raccoons and foxes. When threatened, racers first use their speed to try to escape, often retreating into a bush or tree where they stop and remain motionless. If escape is not possible, racers will not hesitate to defend themselves. A threatened racer generally vibrates its tail and will strike repeatedly at its tormentor. Racers are notorious biters when captured, and because of their quickness usually get in several good bites—often accompanied by intense chewing—before they can be restrained.

CONSERVATION A primary threat to racers, beyond the natural sources of mortality confronting most snakes, is highway mortality. Racers are fast-moving snakes that do not seem to be intimidated by open areas such as fields, grasslands, and highways. Their inclination to cross roads gets many of them killed.

Coachwhips are visual predators with large eyes.

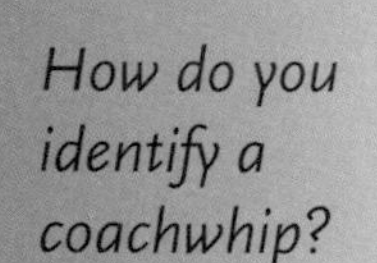

How do you identify a coachwhip?

SCALES
Smooth

ANAL PLATE
Divided

BODY SHAPE
Long and slender

BODY PATTERN AND COLOR
Solid black on head and anterior back and belly, changing to tan on posterior portion of body

DISTINCTIVE CHARACTERS
Tail scales resemble a braided whip

SIZE

BABY
TYPICAL
MAXIMUM
0' 4' 8'

Coachwhip

Masticophis flagellum

DESCRIPTION Adult coachwhips are the only snakes in the Southeast that are solid black anteriorly and completely tan on the posterior portion of the body. The proportion of black to tan varies considerably over the range; those from parts of northern Florida and southern Georgia, for instance, are mostly tan. They have large eyes, a large head, and a tail that resembles a braided whip because of the way the scales are patterned.

WHAT DO THE BABIES LOOK LIKE? Like racers, coachwhip babies do not look at all like the adults. They are light tan with irregular brownish bands along the back and sides.

OTHER NAMES Coachwhips are sometimes called whipsnakes.

DISTRIBUTION AND HABITAT Coachwhips are absent from Virginia and have limited distributions in North Carolina and Tennessee, but are found throughout Florida

Baby coachwhips are light tan with darker bars on the front part of their body.

Coachwhip
Masticophis flagellum

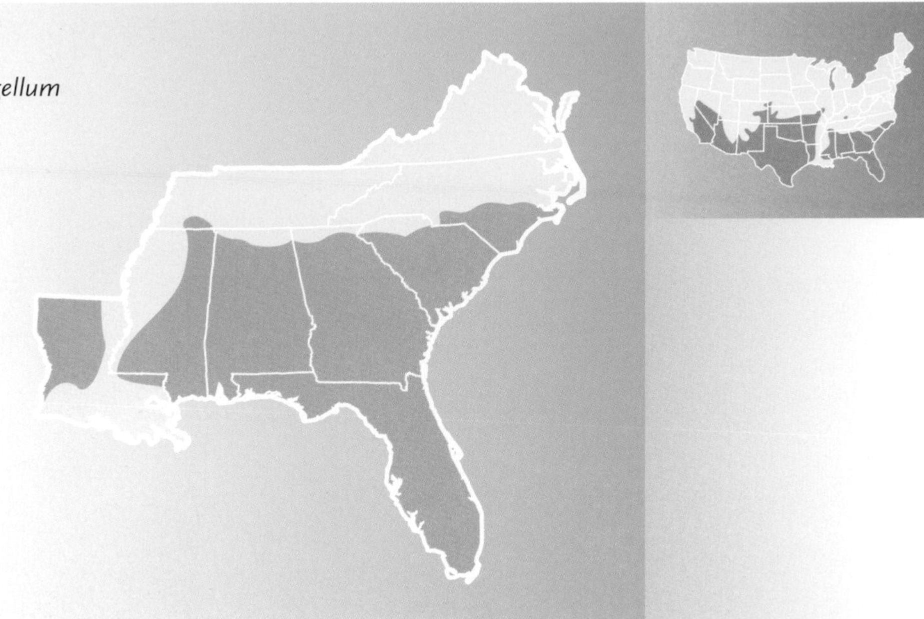

and in large parts of all the other southeastern states. They inhabit open areas such as old fields, longleaf pine forests, palmetto flatwoods, and sandy scrub oak habitats, generally, but not always, in areas with sandy, dry soil. They sometimes climb into shrubs and small trees.

Coachwhips will strike defensively if cornered.

BEHAVIOR AND ACTIVITY Like most other snakes of the Southeast, those living in southern Florida and warm coastal areas are active on some days every month, and those in colder areas hibernate during the winter. Coachwhips generally retreat underground into animal burrows, old root tunnels, or beneath logs or vegetation during cold weather and at night. They are active aboveground only during daylight hours, often in the hottest seasons and at times of day when most other snakes are inactive.

FOOD AND FEEDING Like their relatives the racers, coachwhips actively search for prey during the daytime. They often hunt in open areas, with the head held high above the ground and moving from side to side as they look for the small mammals (e.g., mice, voles, and rabbits), lizards, birds, and other snakes on which they feed. They sometimes climb high into trees or bushes in pursuit of prey. Coachwhips do not constrict their prey,

Coachwhips get their name from the tail, which resembles a braided whip.

Did you know?

In snake species that have male-male combat, males usually get larger than females.

but grab it and swallow it alive. If several small rodents can be captured at once, a coachwhip will hold one or more down with its body while it swallows another.

REPRODUCTION Mating usually occurs in April or May, shortly after hibernation. Courtship has not been documented in this species in the wild, but there is evidence that males fight for females by engaging in a wrestling match in which each male tries to force the other's head to the ground. About 11 eggs (range = 4–24) having a granular appearance like those of racers are laid in June and July. Hatching takes place in the late summer or early fall.

PREDATORS AND DEFENSE The natural predators are limited to larger animals such as coyotes and large hawks. Like racers, coachwhips first use their speed to try and escape any would-be predator, sometimes even climbing into bushes or small trees. If escape is not possible, a threatened coachwhip will strike repeatedly to defend itself and will hold its ground with an open mouth, usually also vibrating its tail. A captured coachwhip will usually try to bite, although some will not bite if handled gently.

CONSERVATION Coachwhips, like racers, inhabit open areas and thus are likely to cross roads. Although coachwhips are among the faster-moving snakes, the odds of being hit by a vehicle steadily increase for those that live near busy highways, and vehicles are probably the greatest threat faced by large adults.

Eastern indigo snakes have an iridescent sheen in sunlight.

Eastern Indigo Snake *Drymarchon corais*

DESCRIPTION Adult indigo snakes are solid black, but their iridescent scales may look gunmetal blue in the sunlight. The reddish to brown coloration on the chin sometimes extends onto the face. They are likely the longest native snake inhabiting the United States.

WHAT DO THE BABIES LOOK LIKE? Indigo snake babies resemble adults, but more of the chin and the area behind the head may be reddish. They may be much lighter in color and speckled, and some have cross-banding or blotching. Compared with most other southeastern snakes, newly hatched indigo snake babies are huge, ranging from 1 to more than 2 feet long.

OTHER NAMES Indigo snakes are known as blue gophers or blue bullsnakes in some areas.

DISTRIBUTION AND HABITAT The historical geographic range in the Southeast included all of Florida and southeastern Georgia and extreme southern Alabama into Mississippi. Today, natural populations in the Southeast are known only from Georgia and Florida. Indigo snakes prefer dry, open habitats with sandy, well-drained soils. They can be found in palmetto stands, open pine forests, sandhills, longleaf pine stands, and turkey oak forests. During warmer months they are often found at the margins of wet areas where they patrol for prey.

How do you identify an indigo snake?

SCALES
Mostly smooth, although males may have a few keeled scales down the back

ANAL PLATE
Single

BODY SHAPE
Powerful, robust

BODY PATTERN AND COLOR
Solid blue-black on back and undersides

DISTINCTIVE CHARACTERS
Chin color usually reddish to brown

SIZE

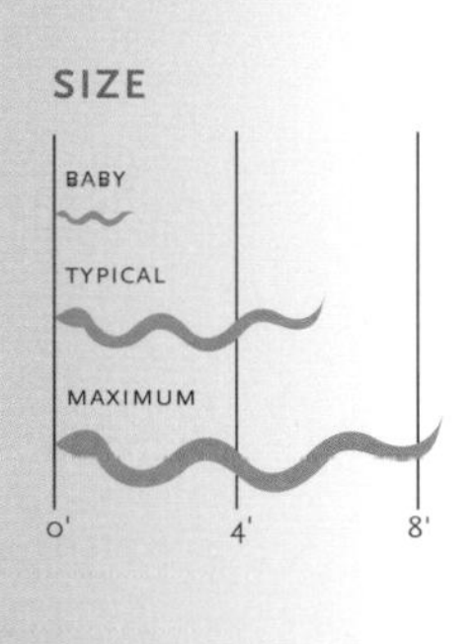

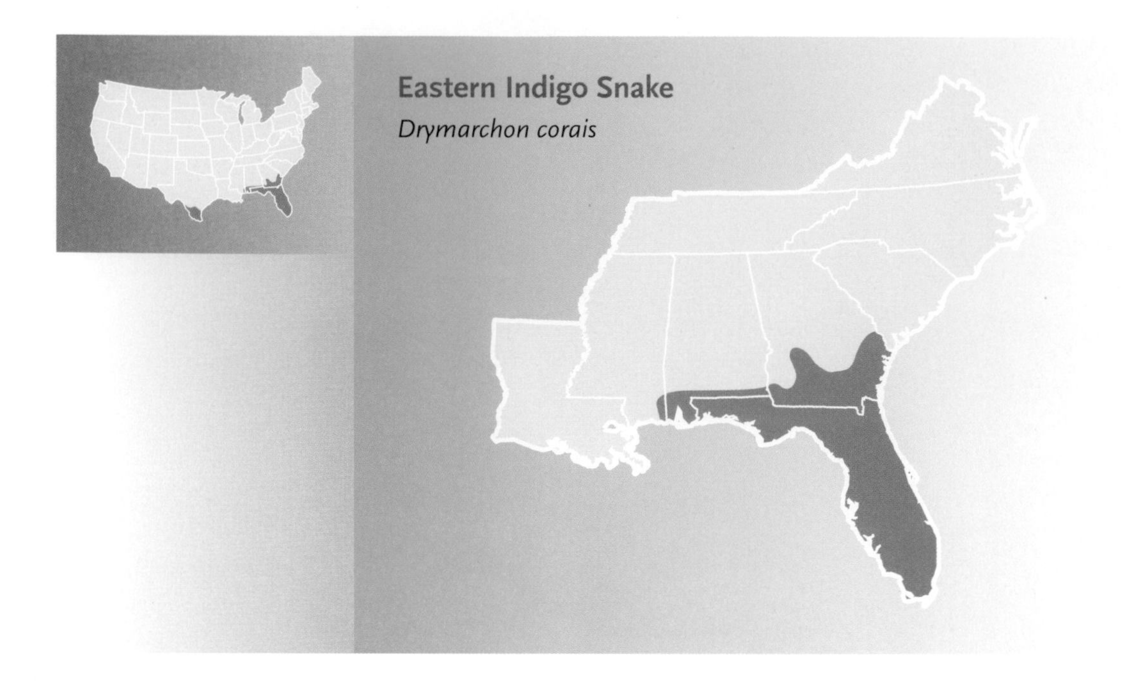

BEHAVIOR AND ACTIVITY Because of their strictly southern geographic range, indigo snakes hibernate for only a few weeks or not at all. Favored retreats are underground burrows of gopher tortoises, armadillos, and other animals. They are active aboveground only during daylight hours and may travel more than 4 miles between their foraging areas and winter retreats.

FOOD AND FEEDING Indigo snakes feed on a wide variety of prey. They are not constrictors, but instead use their size and powerful jaws to overpower their prey. They frequently feed on other snakes, which they grab by the head and pin down with their heavy bodies. Indigo snakes sometimes eat venomous species such as rattlesnakes; if bitten, they may be affected by the venom but usually survive. Other prey animals include lizards, birds and their eggs, rodents and other small mammals, frogs, toads, turtles and their eggs, and even small alligators.

REPRODUCTION Indigo snakes mate from October through March, and possibly through April in Georgia. During courtship the male rubs his chin along the female's back, holding her down until she raises her tail indicating her readiness to mate. Apparently, female indigo snakes can store viable sperm for several years after mating. About 9 eggs (range = 4–14) are laid in May or June, probably in underground burrows made by pocket gophers or gopher tortoises. The extremely large eggs hatch about 3–3.5 months after they are laid.

PREDATORS AND DEFENSE Because of their large size, indigo snakes that reach adulthood have few natural predators. They have been reported to be cannibalistic, so a natural threat to smaller indigo snakes may be larger ones, although individuals observed in the wild usually do not seem to interfere with each other. When threatened, indigo snakes generally try to escape. They seldom bite when first captured or in captivity, but when they do, the strength of their powerful jaws makes it a memorable experience. Indigo snakes sometimes hiss, and unlike other southeastern snakes often flatten the neck vertically when threatened or cornered.

CONSERVATION The eastern indigo snake was officially designated as Threatened under the federal Endangered Species Act in 1978 in the states where it was known to occur (Florida and Georgia) or had been reported historically (Alabama, Mississippi, and South Carolina). Because it is federally protected, the indigo snake is no longer a victim of overcollecting for the pet trade and also is less likely to be a target for deliberate elimination

Eastern indigo snakes often retreat into gopher tortoise burrows.

by people who do not like snakes. Nonetheless, negative public attitudes and local ignorance about snakes (and about the legal implications of the Endangered Species Act) still result in unnecessary killing of these magnificent creatures in some communities within their geographic range. Although not done as overtly as in the past, the illegal practice of pouring gasoline down gopher tortoise burrows during rattlesnake roundups continues in some areas and likely kills any indigo snake remaining in the burrow. One of the greatest threats for a wide-ranging snake species with narrow habitat requirements like the indigo is habitat destruction and degradation resulting from development, fire suppression, and elimination of suitable winter refuges. Fragmentation of suitable habitat by highways, which results in road mortality, is an ever-increasing threat to this species, especially in Florida.

Eastern indigo snakes often have reddish lips and chins.

Vehicular traffic is a potential problem for any snake crossing a highway, and because the eastern indigo snake is one of the longest snakes in the Southeast, road mortality will continue to be a threat. Unfortunately, this species does not appear to have recovered at all since its 1978 federal listing as Threatened, and populations continue to decline or disappear throughout its eastern geographic range.

SCIENTIFIC NOMENCLATURE Herpetologists do not agree on the classification of indigo snakes. Some believe that the eastern indigo should be considered a distinct species (*D. couperi*) separate from the western indigo snake found in south Texas (*D. melanurus*).

WATERSNAKES

The bright red belly color of the black swamp snake has been suggested to be a form of defense against wading birds.

How do you identify a black swamp snake?

SCALES
Smooth

ANAL PLATE
Divided

BODY SHAPE
Moderately stout

BODY PATTERN AND COLOR
Solid black above with red belly

SIZE

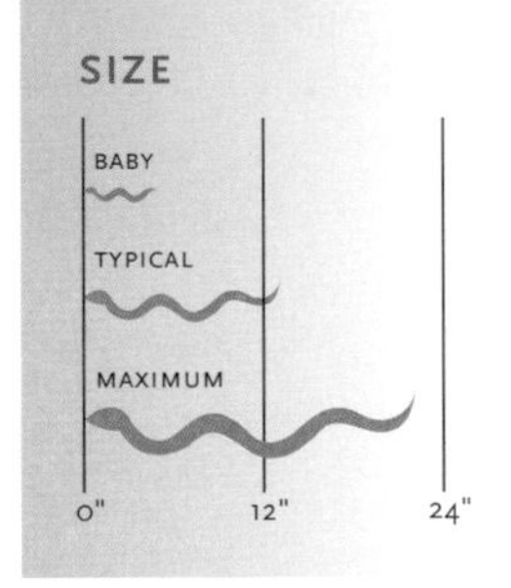

Black Swamp Snake *Seminatrix pygaea*

DESCRIPTION These small, moderately stout snakes have a shiny black back and a red or orange-red belly. The head is relatively small. Some females may be rather heavy bodied at certain times of the year. The smooth scales often have a light line running down the center, giving the impression that they are keeled. The three recognized subspecies (Carolina swamp snake, *S. p. paludis*; North Florida swamp snake, *S. p. pygaea*; and South Florida swamp snake, *S. p. cyclas*) are distinguished primarily by the number of belly scales and the amount of black on each belly scale.

WHAT DO THE BABIES LOOK LIKE? Baby black swamp snakes are identical in appearance to the adults.

DISTRIBUTION AND HABITAT Black swamp snakes are known primarily from Lower Coastal Plain counties from North Carolina to Alabama, most of Florida, and parts of central South Carolina. They are usually associated with still, shallow waters of large or small, heavily vegetated wetlands and generally are not found in or around the moving waters of streams or rivers.

BEHAVIOR AND ACTIVITY Black swamp snakes are characteristically most active in early to late spring but can be found in any month of the year if temperatures are warm enough. In colder regions, they hibernate in the

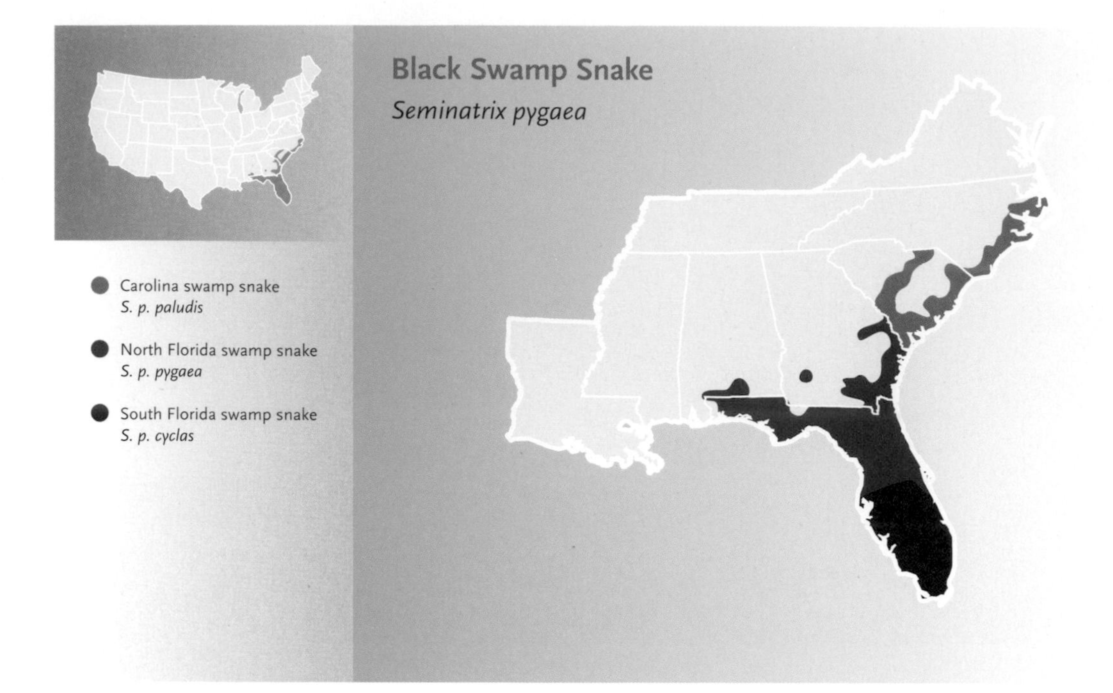

roots of aquatic vegetation, in muskrat lodges, and beneath vegetation or ground litter along the shore. Black swamp snakes are possibly the most aquatic snakes in the Southeast and, unlike the larger watersnakes, seldom bask out of the water. Overland travel occurs during both day and night but may be restricted if conditions are too dry or windy, because they are more susceptible to desiccation than other snakes. Swamp snakes found on land are generally close to shore and usually associated with dense aquatic vegetation.

Black swamp snakes occur in scattered localities but can sometimes be captured in large numbers.

FOOD AND FEEDING Black swamp snakes eat a wide variety of animals found in and around wetland habitats, including leeches, earthworms, small fish, small salamanders and salamander larvae, frogs, and tadpoles. They presumably locate their prey primarily by scent and swallow it alive.

REPRODUCTION Virtually nothing is known about the courtship and mating behavior. Mating probably occurs in late spring, and most females give birth to live young during August or September. Some evidence exists that females emerge onto dry land to give birth. Litter size usually ranges from 2 to 15 (average = around 8) but may be more than 20. Large females have more babies than small ones.

PREDATORS AND DEFENSE Because of their small size, black swamp snakes often fall prey to wetland predators such as turtles, wading birds, raccoons,

and even largemouth bass, as well as to kingsnakes. The bright red belly has been suggested to be a defense mechanism that distracts predatory birds such as herons or egrets that search for prey along the edges of wetlands. The idea is that when the snake is turned over by a bird foraging among the debris along a muddy shoreline, the bright red belly catches the bird's attention; while the bird is looking for a bright red prey item, the shiny black snake turns over and crawls to safety. The red belly has also been suggested to serve as a "warning" signal to predatory animals, although swamp snakes are not venomous. Black swamp snakes rarely bite when picked up.

Black swamp snakes occasionally venture onto land but are primarily aquatic.

South Florida swamp snake from Glades County, Florida

CONSERVATION Their dependence on wetland habitats makes black swamp snakes vulnerable to activities that result in habitat loss, degradation, or isolation.

Juvenile striped crayfish snakes seek out aquatic insect larvae, but adults eat crayfish almost exclusively.

How do you identify a striped crayfish snake?

SCALES
Smooth

ANAL PLATE
Divided

BODY SHAPE
Relatively slender to moderately stout

BODY PATTERN AND COLOR
Alternating brown and yellowish stripes

SIZE

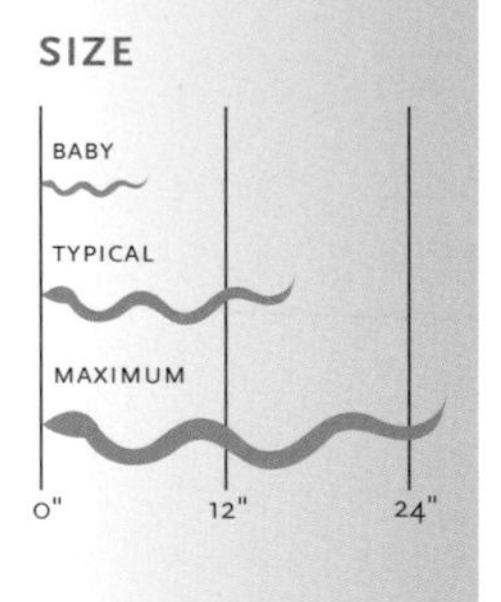

Striped Crayfish Snake *Regina alleni*

DESCRIPTION Striped crayfish snakes generally have three brown stripes running the length of the yellowish brown back. The back often has an iridescent sheen, especially when wet. The belly is cream colored and sometimes has a row of spots running along the center. Striped crayfish snakes are somewhat slender to moderately stout, and the head is not much wider than the neck.

WHAT DO THE BABIES LOOK LIKE? Juvenile striped crayfish snakes look like their parents.

DISTRIBUTION AND HABITAT Striped crayfish snakes are restricted to Florida east of the central panhandle and southeastern Georgia. They are associated with swamps and open wetlands with heavy vegetation and are not usually found around the moving waters of streams or rivers.

BEHAVIOR AND ACTIVITY Striped crayfish snakes are active throughout most of the year except for cold periods during the winter. During active periods they may hide in the roots of aquatic vegetation, including the introduced water hyacinth, or on land beneath logs or debris. They are active in the water during the day and possibly at night, and on cool days bask in sunny spots to warm themselves.

Striped Crayfish Snake

Regina alleni

FOOD AND FEEDING Like other members of the genus *Regina*, striped crayfish snakes specialize on crayfish. Their chisel-like teeth help them to hold on to the hard carapaces of crayfish, which they swallow tail first. There are some reports of them subduing crayfish by constriction before swallowing them. Juveniles eat insect larvae, especially the larvae of dragonflies.

Striped crayfish snakes are primarily aquatic and are active throughout all but the coldest parts of the year.

REPRODUCTION Very little is known about reproduction in this species. They apparently mate in the spring, and young are born in the late summer or early fall. Litter sizes range from about 4 to 12, and larger females produce more babies than smaller females.

PREDATORS AND DEFENSE Predators include great egrets, great blue herons, sandhill cranes, kingsnakes, cottonmouths, large aquatic salamanders, and river otters. When disturbed, striped crayfish snakes try to escape into the water. Individuals release musk from scent glands in the cloaca when captured and may writhe open mouthed in the captor's hand, but most do not bite.

CONSERVATION Because of their specialized diets, striped crayfish snakes can disappear from an area if pollution or other habitat degradation eliminates their primary prey. Therefore, in addition to wetland destruction due to human development, one of the greatest threats to crayfish snakes is habitat alteration that reduces crayfish populations.

Glossy crayfish snakes typically have a cream-colored belly with two rows of spots down the center.

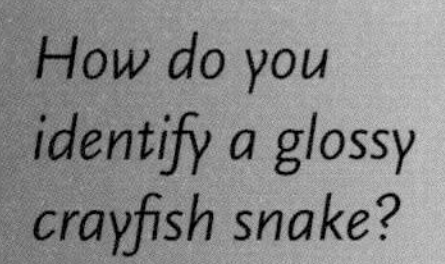

SCALES
Keeled

ANAL PLATE
Divided

BODY SHAPE
Relatively slender to moderately stout

BODY PATTERN AND COLOR
Two light stripes on dark back; belly with two rows of spots

SIZE

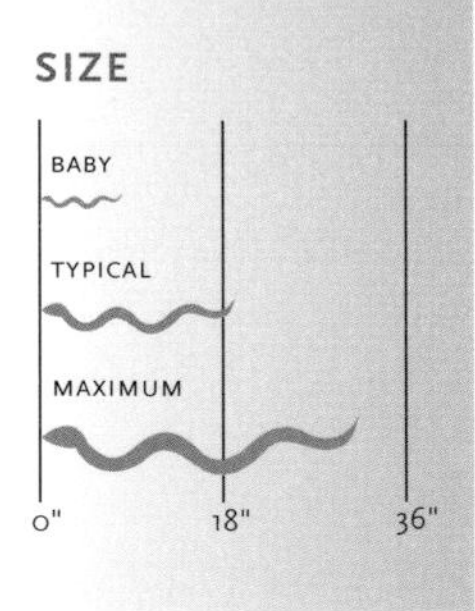

Glossy Crayfish Snake

Regina rigida

DESCRIPTION The body is dark brown with two narrow, light stripes running down the center of the back. The back usually has an iridescent sheen if wet. The cream-colored belly has two rows of spots running down the center. The head is not much wider than the neck. The three subspecies (glossy crayfish snake, *R. r. rigida*; Gulf crayfish snake, *R. r. sinicola*; and Delta crayfish snake, *R. r. deltae*) are distinguished by slight variations in color pattern and scale numbers. Populations with many solid black individuals have been found in the Mississippi Delta and in Florida.

WHAT DO THE BABIES LOOK LIKE? Juvenile glossy crayfish snakes look like their parents, but their bellies may be somewhat pink.

DISTRIBUTION AND HABITAT Glossy crayfish snakes are spottily distributed from Virginia into the southeastern halves of the Carolinas; in parts of central and northern Florida; in the southern halves of Georgia, Alabama, and Mississippi; and in most of Louisiana in swamps, floodplains, and open wetlands with heavy vegetation. They are not usually found around swift-moving streams or rivers.

BEHAVIOR AND ACTIVITY Glossy crayfish snakes are active throughout most of the year except for cold periods during the winter, when they usually hibernate in crayfish burrows. They are apparently active mostly at night, although behavior in this species has rarely been documented. They can sometimes be found basking on tree limbs in river swamps on sunny days during late winter and early spring.

FOOD AND FEEDING Like other members of the genus *Regina*, glossy crayfish snakes specialize on crayfish. Their chisel-like teeth fit into the grooves of crayfish shells, providing a secure hold for the snake. They may constrict their prey and always swallow it tail first. Other reported prey items include small frogs and fish. Juveniles eat insect larvae, especially the larvae of dragonflies.

The Delta crayfish snake is a subspecies of glossy crayfish snake restricted primarily to the Mississippi River delta of Mississippi and Louisiana.

REPRODUCTION Very little is known about the reproduction of glossy crayfish snakes. They apparently mate in the spring, and young are born in the late summer or early fall. Litters can range from about 6 to 14 babies, and larger females probably produce more babies than smaller females.

PREDATORS AND DEFENSE Because these snakes are so rare, documentation of predators is difficult. They likely fall victim to any larger vertebrate that might prey on a snake of their size, including large aquatic

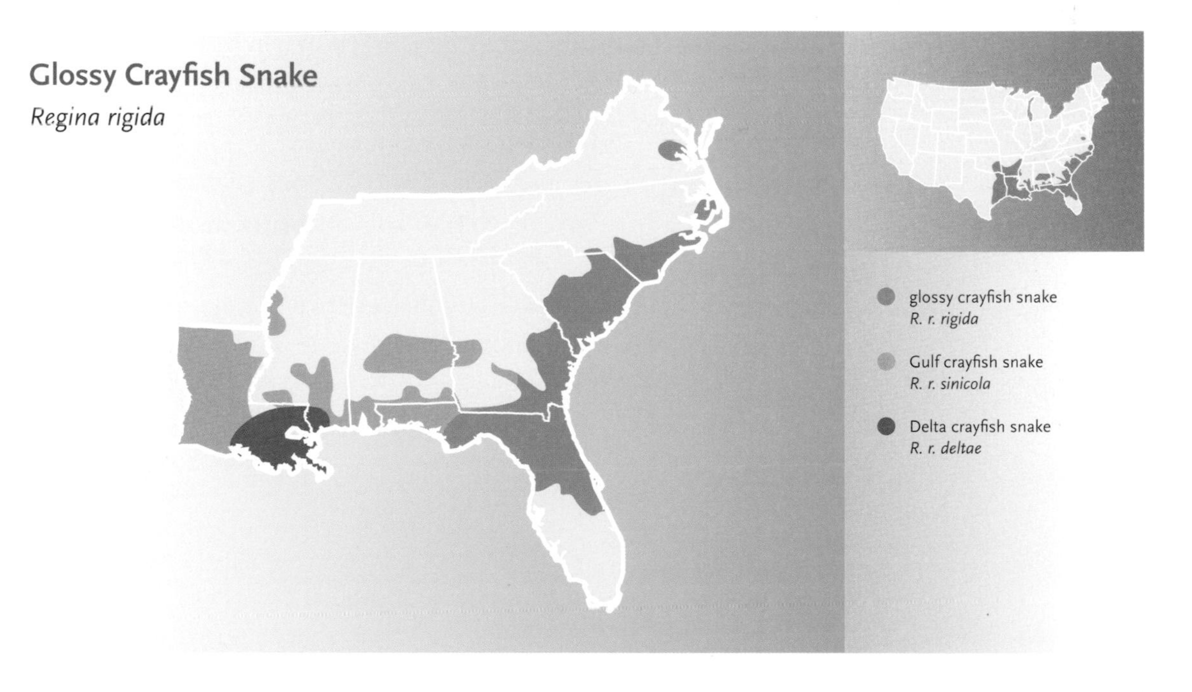

The Gulf crayfish snake, a subspecies of the glossy crayfish snake, ranges from Georgia to western Louisiana.

Glossy crayfish snake from Aiken County, South Carolina

Did you know?

A giant salamander (the amphiuma) has been documented as a predator of the glossy crayfish snake.

salamanders, cottonmouths, and kingsnakes. When disturbed, they try to escape into the water. If captured they release musk from their cloacal glands but rarely, if ever, bite.

CONSERVATION Records of glossy crayfish snakes are too scarce to determine their status in most regions. They are particularly rare in North Carolina and Virginia, and because there are so few locations where populations are known to occur, they should be considered species of special concern. Pollution or other forms of habitat degradation that eliminate the crayfish that are their primary prey will result in the loss of this species as well.

In the Southeast, Graham's crayfish snakes are known only from widely separated localities in Mississippi and Louisiana.

Graham's Crayfish Snake *Regina grahamii*

DESCRIPTION Graham's crayfish snakes are brown or grayish brown with a broad yellow stripe running the length of each side and, usually, a dark stripe running down the middle of the back. They are generally slender, but some large females are somewhat stocky. The head is not much wider than the neck. The belly is solid yellow, sometimes with a central row of spots.

WHAT DO THE BABIES LOOK LIKE? Baby Graham's crayfish snakes look like the adults but may be more boldly marked.

DISTRIBUTION AND HABITAT This species has one of the patchiest geographic distributions of any southeastern snake. In the Southeast it is known only from scattered localities in Mississippi and Louisiana, where it is associated with a variety of wetland types, including small lakes and ponds, swamps, bayous, and slow-moving streams where crayfish abound.

BEHAVIOR AND ACTIVITY Graham's crayfish snakes hibernate in colder regions, commonly using crayfish burrows as winter refuges. In spring and fall they are most active during the day, but in summer they are occasionally found at night as well. Individuals commonly bask during the day on tree limbs, bushes, and other vegetation along the edges of aquatic habitats.

How do you identify a Graham's crayfish snake?

SCALES
Keeled

ANAL PLATE
Divided

BODY SHAPE
Slender to moderately stout

BODY PATTERN AND COLOR
Brown with a broad yellowish stripe on each side

SIZE

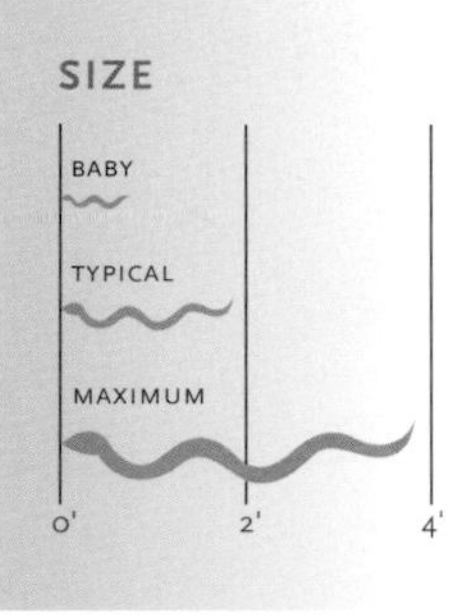

Graham's crayfish snakes typically have a yellow belly with a central row of dark spots.

They move overland between aquatic habitats and are sometimes found on land along shorelines.

FOOD AND FEEDING As the name implies, Graham's crayfish snakes eat primarily crayfish, specializing on newly molted crayfish (those with a recently shed exoskeleton) with soft shells. They occasionally eat small frogs and minnows as well. They do not kill their prey, but simply swallow it alive.

REPRODUCTION Graham's crayfish snakes mate in the spring (April and May) and give birth to live young in the late summer or early fall. Little is known about their courtship, but on several occasions

Graham's Crayfish Snake

Regina grahamii

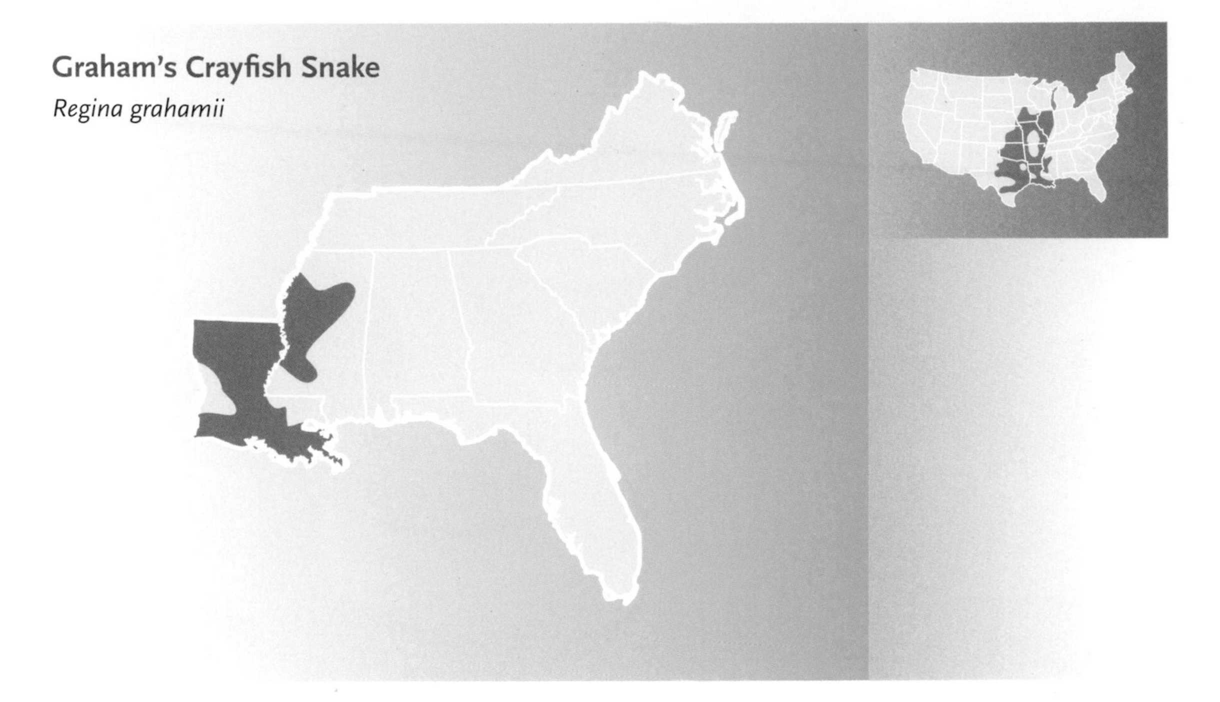

multiple males have been observed trying to mate with one female. Litter size ranges from 9 to 39, with an average of about 15. During gestation, females apparently can transfer nutrients to the developing offspring much as mammals do, in contrast to many live-bearing snakes in which the embryos acquire their nourishment only from an attached yolk sac.

PREDATORS AND DEFENSE Natural predators associated with wetlands, such as wading birds, large fish, and snake-eating snakes, will eat this snake. Graham's crayfish snakes are generally mild-mannered when captured but usually release an ill-smelling musk from glands at the base of the tail.

CONSERVATION Their strict dependence on healthy crayfish populations makes this species particularly vulnerable to various forms of habitat degradation.

A brown body with a broad yellowish stripe on each side is characteristic of Graham's crayfish snake.

The dark stripes running the length of the back of queen snakes are more conspicuous in some specimens than in others.

Queen Snake

Regina septemvittata

How do you identify a queen snake?

SCALES
Keeled

ANAL PLATE
Divided

BODY SHAPE
Slender to moderately stocky

BODY PATTERN AND COLOR
Brown, grayish, or olive green above with yellowish stripes on the sides

SIZE

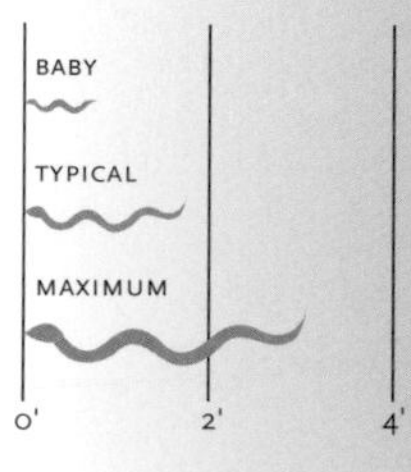

DESCRIPTION Queen snakes have a light brown, grayish, or olive green back with three darker, usually inconspicuous brown stripes running along the entire length. A yellow stripe runs down each side, and the belly is yellowish with four brown stripes running its length. Queen snakes are slender, although females may be more robust. The head is not much wider than the neck.

WHAT DO THE BABIES LOOK LIKE? Baby queen snakes look like their parents.

DISTRIBUTION AND HABITAT Queen snakes are found in every southeastern state except Louisiana, but often in widely separated areas. They are absent from most of the southeastern halves of the Carolinas and Georgia, northern Mississippi, the western quarter of Tennessee, and all of peninsular Florida. These are the quintessential water snakes of cold, clear, rocky- or sandy-bottomed streams lined with shrubby vegetation and where crayfish are abundant. They also can be found in heavily vegetated backwaters of rivers, lakes, and ponds, and sometimes in reservoirs.

BEHAVIOR AND ACTIVITY Queen snakes become inactive during the winter in most parts of the Southeast, retreating inside root masses, under vegetation and ground litter near shore, or into burrows made by other animals along streams. During warmer months, queen snakes can be found under large rocks or logs, partially underwater. They commonly bask on the limbs of shrubs and small trees overhanging the water in spring and fall, and occasionally bask during the summer as well. They are active primarily during the day.

FOOD AND FEEDING Queen snakes, like Graham's crayfish snakes, their close relatives, feed almost exclusively on defenseless, soft-shelled, newly molted crayfish. They avoid hard-shelled crayfish that can defend themselves with large pincers. In addition to crayfish, queen snakes sometimes eat small fish. All prey animals are grabbed and swallowed alive.

REPRODUCTION Like other members of the genus *Regina*, queen snakes apparently mate in the spring, although at least one observation indicates that they may have a fall mating season as well. Queen snakes give birth to 5–23 (average = about 11) live young during summer or early fall.

PREDATORS AND DEFENSE Natural predators of queen snakes are primarily those associated with stream habitats or their margins and include racers, kingsnakes, cottonmouths, large fish, hellbender salamanders, and, ironi-

Queen snakes have four stripes down the belly, two near the center and two on the sides.

Queen Snake
Regina septemvittata

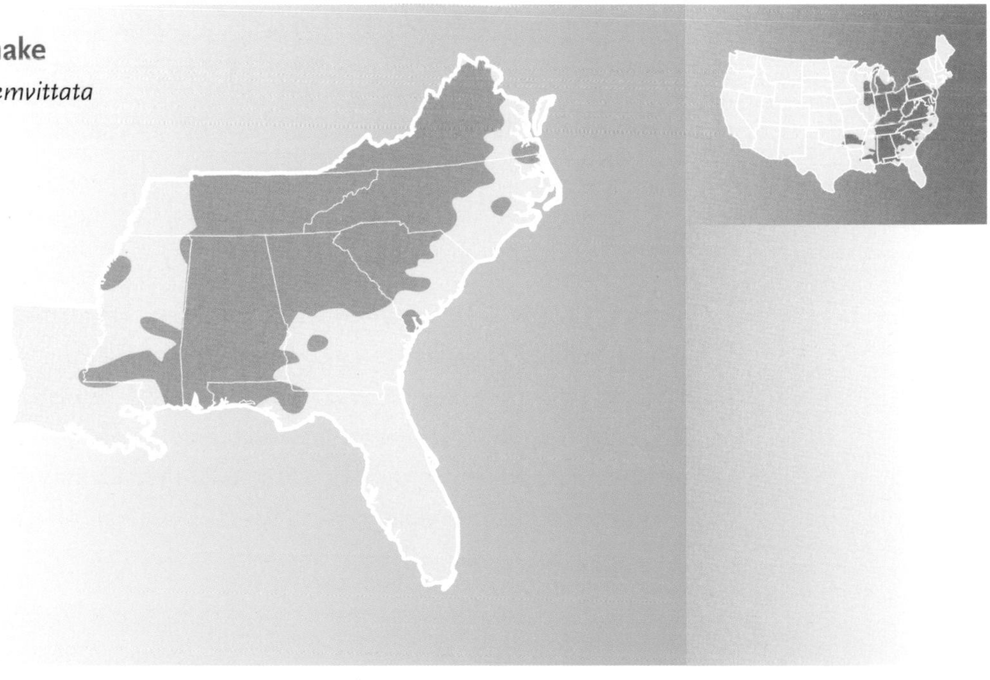

Queen snakes have proportionately smaller heads when compared with the larger watersnakes.

Queen snake from Okaloosa County, Florida

cally, perhaps even large crayfish. Queen snakes writhe and release musk when captured but rarely bite.

CONSERVATION Like the three other species of crayfish-eating snakes, queen snakes have a vital relationship with the environmental health of their habitat, apparently because of their dietary dependence on crayfish. Thus, primary threats to queen snakes include stream pollution, siltation, stream channelization, and other forms of degradation that reduce or eliminate crayfish populations.

Northern watersnakes commonly bask throughout their geographic range.

How do you identify a northern watersnake?

SCALES
Keeled

ANAL PLATE
Divided

BODY SHAPE
Heavy bodied

BODY PATTERN AND COLOR
Variable; usually dark brown blotches alternating with blotches on sides

SIZE

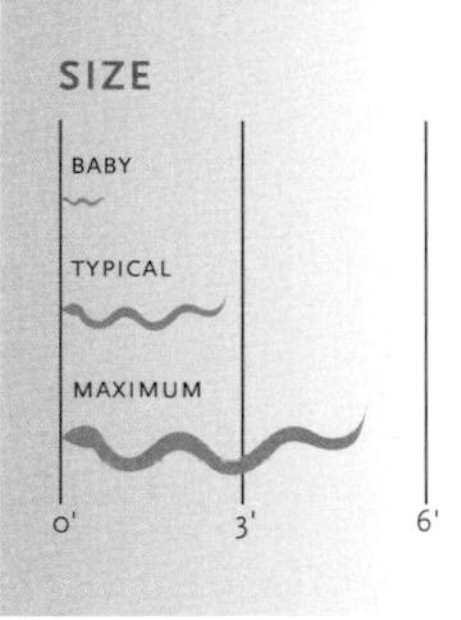

Northern Watersnake *Nerodia sipedon*

DESCRIPTION Both coloration and pattern are highly variable. Typically, northern watersnakes have alternating bands of reddish brown and lighter brown on the front part of the body, changing to alternating square blotches on the back and sides toward the rear of the body. Large adults are usually solid dark brown or even dull black. The belly is typically yellowish or cream colored with a half-moon spot on each belly scale. Three subspecies occur in the Southeast. The northern watersnake (*N. s. sipedon*), which has 30 or more body bands or blotches, and the midland watersnake (*N. s. pleuralis*), which has fewer than 30, are both wide-ranging subspecies. The Carolina watersnake (*N. s. williamengelsi*), found only on the Outer Banks of North Carolina and the adjacent mainland, is typically much darker than the other two forms.

WHAT DO THE BABIES LOOK LIKE? Babies have greater contrast between dark and light crossbands and blotches.

OTHER NAMES In the South this species is traditionally known as the banded watersnake both because of its appearance and because it was once considered the same species as the southern banded watersnake.

DISTRIBUTION AND HABITAT Northern watersnakes are found throughout Virginia and most of North Carolina south through the northwestern halves of Alabama and Georgia, most of Alabama and Tennessee, and sporadically throughout Mississippi. This species and its southern counterpart, the southern banded watersnake, are likely to be found in any aquatic habitat within their geographic range, including suitable habitat on barrier islands. Fish hatcheries; small streams; large rivers; swamps; marshes; Carolina bay wetlands; bayous; and all ponds, lakes, and reservoirs are home to this species.

Northern watersnake from Davidson, North Carolina

BEHAVIOR AND ACTIVITY Northern watersnakes hibernate during the winter and during the warmer months retreat beneath a variety of hiding places such as root masses; burrows of other animals; overhanging shorelines; and near-shore vegetation, logs, and rocks. In northern regions individuals have been observed traveling overland to hibernation sites several hundred feet from water, and some southeastern populations may do the same. During sunny days in the spring and fall, and cooler periods during the summer, northern watersnakes in the Southeast bask on trees and shrubs along the shore, always with a ready escape route to the water. They may be active during the daytime in spring and fall but are more likely to be active in the water at night during warm

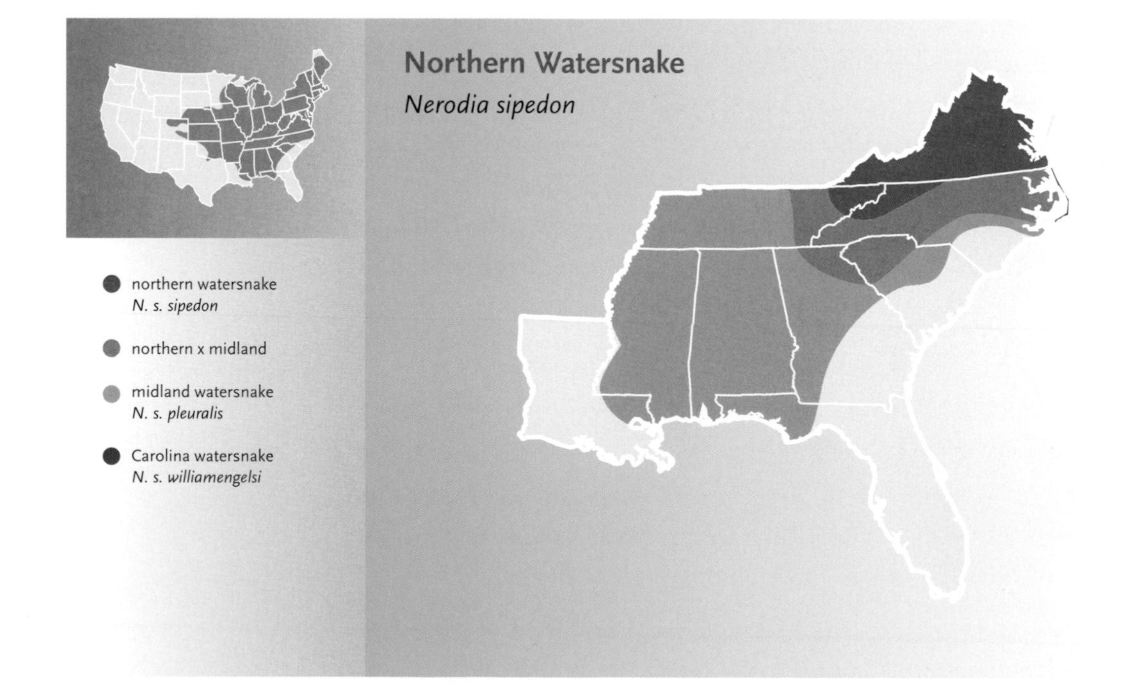

Midland watersnake, a subspecies of the northern watersnake, from Horsehoe Bend National Military Park, Alabama

The Carolina watersnake is generally darker than the other subspecies of northern watersnakes.

periods of the year. Northern watersnakes in some areas make frequent journeys overland between aquatic habitats.

FOOD AND FEEDING Northern watersnakes eat a greater variety of fish and amphibians than any other North American watersnake; more than 80 fish species and 30 amphibians have been recorded as prey. Although they eat mostly vertebrates, invertebrates such as insects and leeches are eaten when the opportunity arises. In addition to being active foragers, northern watersnakes will wedge themselves among rocks and move back and forth with the mouth open ready to catch unsuspecting fish, and they sometimes corral fish in shallow water with a body coil. All prey items are swallowed alive.

REPRODUCTION Northern watersnakes mate during late April, May, and early June. It is not uncommon for many males to try simultaneously to

Northern watersnakes typically have bands on the front part of the body that change to alternating square blotches toward the rear of the body.

Northern waternakes typically have half-moon-shaped spots on each belly scale.

mate with a single, larger female. Females sometimes mate with more than one male and bear litters with multiple fathers. Litter sizes range from 4 to nearly 100, but a typical number is 20–30. The embryos are nourished by the mother via a placenta-like structure during development, and birth takes place in late summer or early fall.

The banding pattern of northern watersnakes can become obscured in older specimens or prior to shedding.

PREDATORS AND DEFENSE Adults and juveniles are taken by a wide variety of natural predators, including many species of birds, mammals, turtles, other snakes, and predatory fish. A threatened watersnake will try to escape, but if that is not possible it will spread its jaws slightly and flatten the front part of the body to make the head look bigger and will strike and bite to defend itself. If picked up, northern watersnakes characteristically writhe, bite, and release musk from their anal glands.

CONSERVATION Northern watersnakes are common over most of their range because they can live in a variety of aquatic habitats. Unfortunately, they are frequent victims of anglers and wildlife agents who mistakenly believe that they prey on trout and other game fish at a level that affects fishing success. And each year, many harmless northern watersnakes are mistaken for venomous cottonmouths and killed, even in areas hundreds of miles from where cottonmouths live.

Southern banded watersnakes are often mistaken for venomous species because of their large heads and superficial similarity to copperheads or cottonmouths.

Southern Banded Watersnake *Nerodia fasciata*

DESCRIPTION Southern banded watersnakes are stout snakes that typically have irregular dark brown or reddish brown bands on a lighter background. Large adults sometimes become solid dark brown or dull black. The belly is yellowish, and there are red spots or wavy markings on each belly scale. All three subspecies (banded watersnake, *N. f. fasciata*; Florida watersnake, *N. f. pictiventris*; and broad-banded watersnake, *N. f. confluens*) occur in the Southeast. The broad-banded watersnake has fewer (about 10–11) and much wider bands than the other two subspecies.

WHAT DO THE BABIES LOOK LIKE? Babies have banding similar to that of the adults but with sharper contrast between dark bands and background coloration.

DISTRIBUTION AND HABITAT Southern banded watersnakes occur in almost every aquatic habitat in the Coastal Plain of the Carolinas and Georgia, throughout Florida, in the southern portions of Alabama and Mississippi, throughout Louisiana, and up the Mississippi River drainage.

BEHAVIOR AND ACTIVITY Banded watersnakes are active year-round in the southern parts of their range and become inactive during cold periods in the winter in the other areas. They hibernate in animal burrows near the water,

How do you identify a southern banded watersnake?

SCALES
Keeled

ANAL PLATE
Divided

BODY SHAPE
Heavy bodied

BODY PATTERN AND COLOR
Dark crossbands on a light brown body

DISTINCTIVE CHARACTERS
Dark brown or black stripe behind eye and onto neck

SIZE

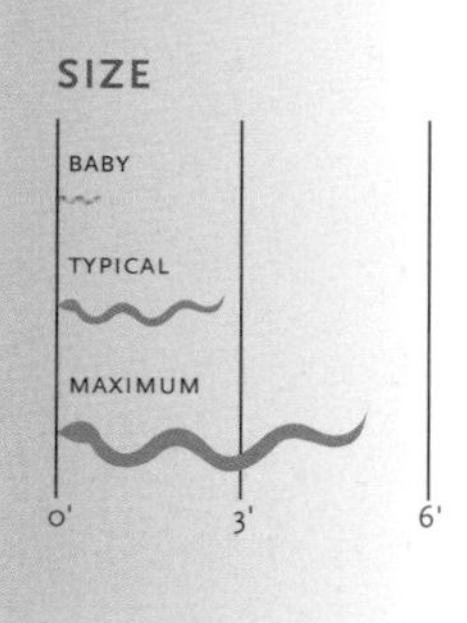

muskrat and beaver lodges, fallen logs, and shoreline vegetation, and they retreat to the same places during warmer weather. They are active at night in many areas during the summer but also may be active during the day. On cool, sunny days they bask on limbs overhanging rivers, streams, ponds, and other aquatic habitats.

The broad-banded watersnake is a western subspecies found throughout Louisiana.

FOOD AND FEEDING Banded watersnakes are active foragers that eat a variety of aquatic animals, mostly fish and amphibians. Most fish are swallowed head first, and most frogs are swallowed rear first. Prey are not subdued but are swallowed whole while still alive.

REPRODUCTION Like other North American watersnakes, banded watersnakes typically mate during the spring and give birth to live young from midsummer into fall. Litters range from 6 to as many as 80 young but typically number about 20–25. Multiple males sometimes congregate in shallow-water habitats during the early spring to mate with receptive females.

Large southern banded watersnakes are often dark with no visible bands.

Southern Banded Watersnake

Nerodia fasciata

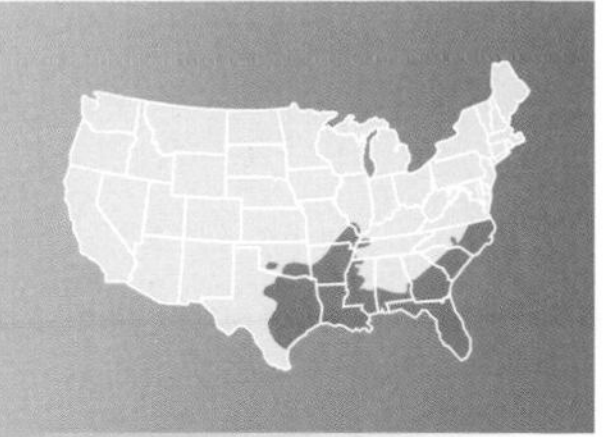

banded watersnake
N. f. fasciata

Florida watersnake
N. f. pictiventris

broad-banded watersnake
N. f. confluens

The Florida watersnake, a subspecies of the southern banded watersnake, is found throughout peninsular Florida.

Did you know?

Sometimes snakes congregate around a particularly suitable habitat, but unless they are mating, they do so because of the environmental conditions, not because other snakes are there.

Banded watersnake from Berkeley County, South Carolina

In some southern banded watersnakes, the characteristic stripe from the eye to the corner of the jaw can be obscure.

PREDATORS AND DEFENSE Cottonmouths, great blue herons, and alligators are all known predators of banded watersnakes, as are numerous other predators that live in and around aquatic ecosystems within their range. A threatened banded watersnake always tries to escape first, but when captured will twist around, bite, and release musk.

CONSERVATION Like other watersnakes, banded watersnakes are intentionally killed by campers, fishermen, and boaters in many regions. They are often mistaken for copperheads because of the banded color pattern or for cottonmouths because of their presence around water. An additional threat to banded watersnakes is their tendency to cross highways that are adjacent to their wetland habitats.

The belly of the southern banded watersnake is yellowish with red markings.

The Gulf salt marsh snake characteristically has four dark longitudinal stripes on the back and sides.

Salt Marsh Snake *Nerodia clarkii*

DESCRIPTION Salt marsh snakes exhibit a bewildering variety of color patterns. The back and sides may be solid, striped, or blotched, and the belly may be solid or spotted. Body colors include brown, black, yellow, white, and red in various combinations above and below. Three subspecies are recognized. The Gulf salt marsh snake (*N. c. clarkii*) is the least variable in color pattern, usually having four longitudinal stripes on a lighter background. The mangrove salt marsh snake (*N. c. compressicauda*) may be solid, banded, blotched, or even somewhat striped. The Atlantic salt marsh snake (*N. c. taeniata*) is also highly variable but usually has stripes on the first one-third of its body. Intergrades occur in the area where the ranges of the Gulf and mangrove salt marsh snake overlap (see map).

WHAT DO THE BABIES LOOK LIKE? Baby salt marsh snakes are as variable as the adults, but their stripes and spots are usually more distinctive. Mangrove salt marsh snakes can have striped, solid, and blotched babies in a single litter.

DISTRIBUTION AND HABITAT The salt marsh snake is unique among North American snakes in being partial to brackish water and inhabiting the coastal margins in each of the Gulf Coast states. The only Atlantic Coast populations are in Florida. The species is commonly associated with salt

How do you identify a salt marsh snake?

SCALES
Keeled

ANAL PLATE
Divided

BODY SHAPE
Moderately stout to heavy bodied

BODY PATTERN AND COLOR
Extremely variable: banded, solid, or striped

SIZE

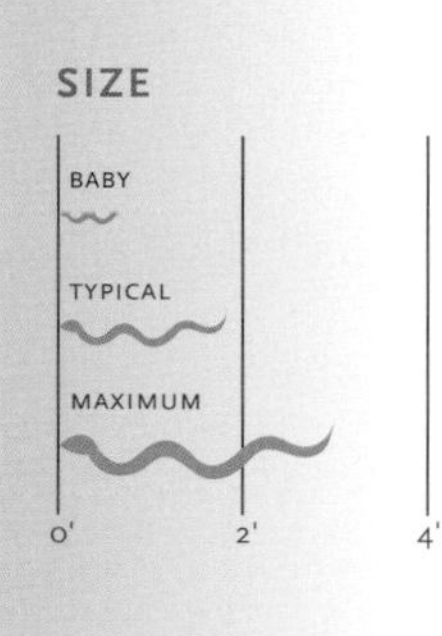

The Atlantic salt marsh snake is a federally threatened species.

marsh and mangrove swamps, but individuals will occasionally enter saltwater areas that are contiguous with freshwater habitats and enter fresh water as well when it is next to brackish water.

BEHAVIOR AND ACTIVITY Salt marsh snakes are active year-round in southern Florida but are inactive for short intervals during winter along most of the Gulf Coast. They bask less frequently than the other large watersnakes and are usually inactive during the day, hiding in vegetation, in animal burrows (including those made by crabs), and under logs or marsh grass. They forage in the water, mainly at night, and their activity is greatly influenced by tidal cycles in some areas.

FOOD AND FEEDING Salt marsh snakes primarily eat small fish that live in brackish water, which they may lure by wiggling and curling the tongue in the water and at the water's surface. They occasionally eat crustaceans such as fiddler crabs.

REPRODUCTION The reproductive biology is poorly known, although they apparently mate during the spring and give birth in late summer and early fall. Litter size ranges from 1 to 24 with an average of about 10.

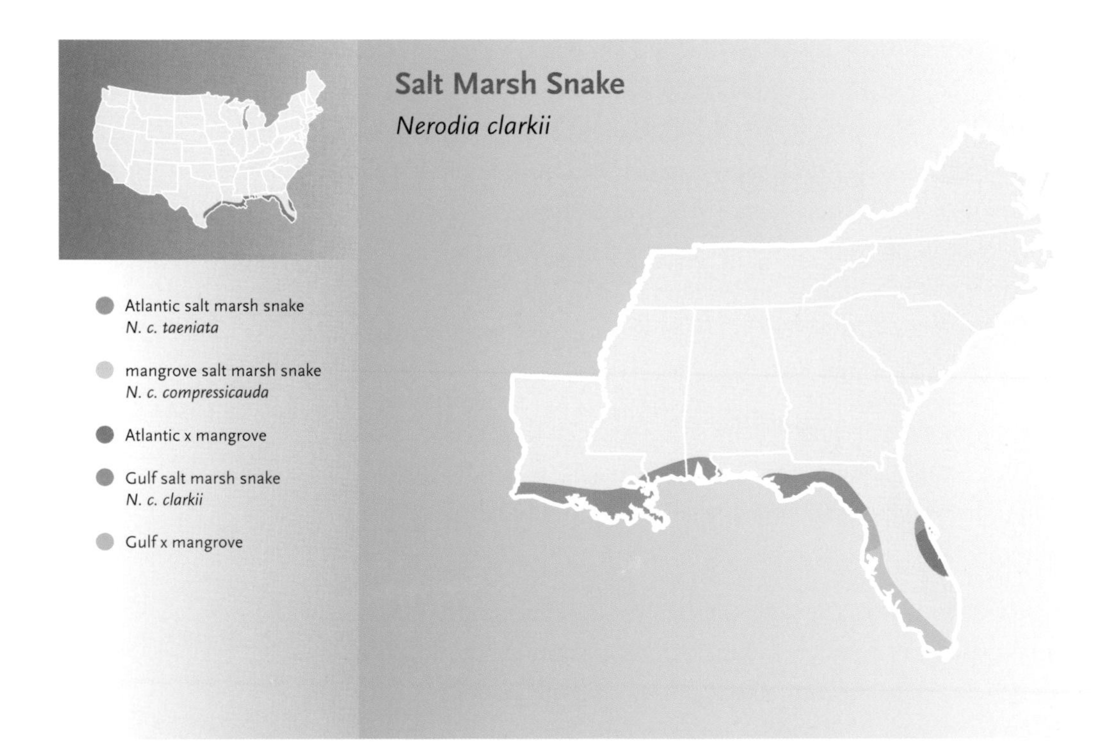

PREDATORS AND DEFENSE Natural predators of salt marsh snakes are those associated with brackish water habitats and include marine fish, herons and egrets, and large crabs. They rely on camouflage for protection and react more mildly to capture than the larger watersnakes, sometimes not even biting.

CONSERVATION In 1977, the subspecies known as the Atlantic salt marsh snake was officially designated as Threatened throughout its entire range (Volusia County, Florida) under the federal Endangered Species Act. Deterioration of natural barriers that occurs when canals are constructed between brackish and freshwater habitats can result in interbreeding between salt marsh snakes and their close relatives, banded watersnakes. A population that interbreeds with banded watersnakes will lose its genetic integrity. Most populations may be imperiled in the near future as a result of coastal development.

Mangrove salt marsh snakes from Florida are highly variable in color and body pattern.

Red-bellied watersnakes are at home in most wetlands but frequently travel overland between aquatic areas.

How do you identify a plain-bellied watersnake?

SCALES
Keeled

ANAL PLATE
Divided

BODY SHAPE
Heavy bodied

BODY PATTERN AND COLOR
Usually a solid brown back and plain yellow to red belly

SIZE

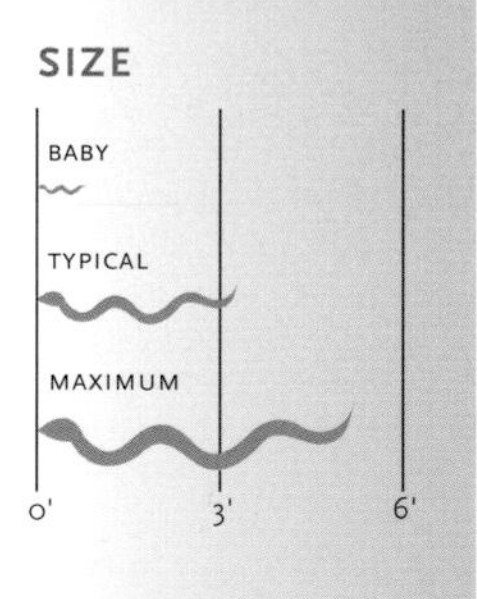

Plain-bellied Watersnake *Nerodia erythrogaster*

DESCRIPTION Plain-bellied watersnakes are stout snakes with a brown, gray, or greenish gray back and, as the common name implies, a plain, unmarked belly ranging in color from yellow to red. Three subspecies occur in the Southeast: the red-bellied watersnake (*N. e. erythrogaster*) has a red to orange belly; the yellow-bellied watersnake (*N. e. flavigaster*) has a light orange to yellow belly; and the blotched watersnake (*N. e. transversa*) has a yellow belly and a faded pattern of alternating blotches on its back.

WHAT DO THE BABIES LOOK LIKE? Babies have banding patterns that closely resemble those of banded watersnakes, but can be identified by their unmarked bellies and bands that are incomplete in the neck region.

Baby plain-bellied watersnakes are banded, but the bands are incomplete in the neck region.

A plain yellow, orange, or red belly is characteristic of red-bellied watersnakes.

DISTRIBUTION AND HABITAT Plain-bellied watersnakes are found in every southeastern state from southeastern Virginia and eastern North Carolina west to western Tennessee and south to northern Florida. Because of their propensity to travel long distances overland, plain-bellied watersnakes can turn up in or around any aquatic habitat. They are commonly associated with rivers and floodplains, large and small lakes and ponds, and other natural wetlands. They are equally likely to be found in clear, fast-moving streams, in sluggish backwaters of river swamps, and in lakes and ponds.

This young red-bellied watersnake is basking on a limb overhanging the water.

BEHAVIOR AND ACTIVITY Plain-bellied watersnakes hibernate during the coldest part of the winter but frequently become active earlier in the spring than other watersnakes in their area. They can be found throughout the warmer months basking on vegetation along rivers, streams, and open bodies of water; swimming in aquatic habitats; or traveling overland. They

are more likely to be found on land than any other watersnake. They are active during the day in most seasons and at night during the summer months.

FOOD AND FEEDING Plain-bellied watersnakes eat primarily frogs, toads, and fish. Because they travel overland more than other watersnakes, they probably encounter and eat more amphibians than fish. They frequently actively forage for prey but have been known to sit in the water open mouthed, waiting for prey to come near. Like other watersnakes, they capture and swallow prey alive without using constriction.

The blotched watersnake is a western subspecies of the plain-bellied watersnake that enters southwestern Louisiana.

REPRODUCTION Mating occurs from April until mid-June. Females give birth to live young, usually during August or September. Although excep-

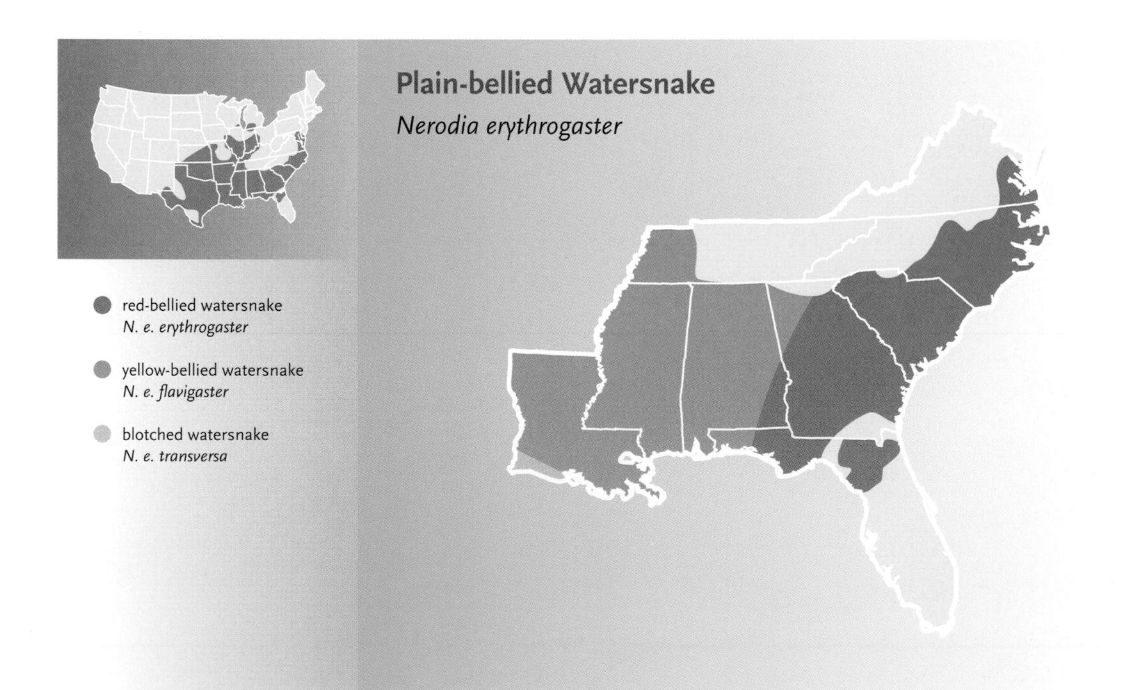

Red-bellied watersnake from Aiken County, South Carolina

Like other large watersnakes, plain-bellied watersnakes will expand the head in a threat display.

tionally large litters have been reported (e.g., 55 babies from one female in North Carolina), a typical litter size is about 18 (range = 2–55).

PREDATORS AND DEFENSE Plain-bellied watersnakes fall prey to both terrestrial and aquatic predators because of their tendency to move overland between aquatic habitats. Documented predators include largemouth bass, kingsnakes, cottonmouths, egrets, red-shouldered hawks, and red-tailed hawks. They typically react to threats like the other large watersnakes—by trying first to escape and then biting and releasing musk if the threat persists. In contrast to other watersnakes, though, this species will often leave the water and attempt to escape on land.

A yellow or white belly is characteristic of the yellow-bellied watersnake.

CONSERVATION Although federal legislation does not protect the plain-bellied watersnake in any of the southeastern states, in 1997 the northern subspecies known as the copperbelly watersnake was officially designated as Threatened in Michigan, Ohio, and northern Indiana under the federal Endangered Species Act. Their propensity for overland movement increases their chances of encountering highways and subsequent road mortality. Wetland degradation can also threaten this species, as individuals often depend on more than one aquatic habitat during the active season.

Diamondback watersnakes vary little in pattern or color throughout their southeastern geographic range.

How do you identify a diamondback watersnake?

SCALES
Keeled

ANAL PLATE
Divided

BODY SHAPE
Heavy bodied, broad head

BODY PATTERN AND COLOR
Brown to grayish green with diamond-shaped blotches

DISTINCTIVE CHARACTERS
Eyes are set more on top of the head than on the sides like those of most other snakes

SIZE

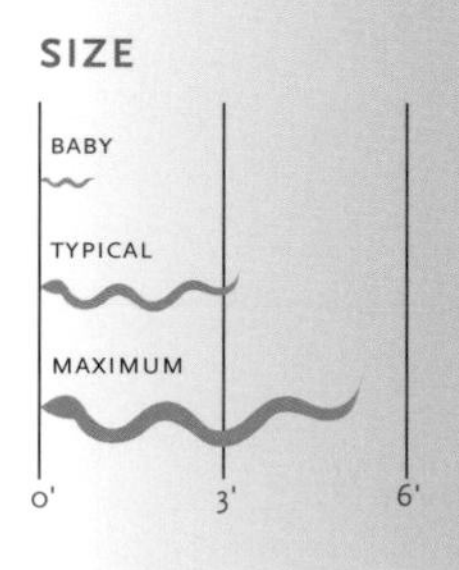

Diamondback Watersnake *Nerodia rhombifer*

DESCRIPTION Diamondback watersnakes are stout to very stout snakes with a series of somewhat diamond-shaped blotches on the back alternating with smaller blotches on the sides. The background color is usually brown, gray, or olive-gray. The blotches typically form a light, chainlike pattern running the length of the snake. The belly is usually cream to yellow with brown or black spots.

WHAT DO THE BABIES LOOK LIKE? The babies look the same as adults but are more boldly marked and often have orangish bellies with black spotting.

DISTRIBUTION AND HABITAT In the Southeast, diamondback watersnakes range from central Alabama and the western tip of the Florida panhandle through Mississippi and Louisiana and parts of the western portion of Tennessee. They can turn up in large numbers in almost any habitat with standing or moving water, including small ponds, rivers, large lakes, and reservoirs. They are frequently seen around spillways of lakes, presumably in search of the small fish that abound in such areas.

BEHAVIOR AND ACTIVITY Diamondback watersnakes are generally inactive in the winter but may become temporarily active during warm spells.

Diamondback Watersnake

Nerodia rhombifer

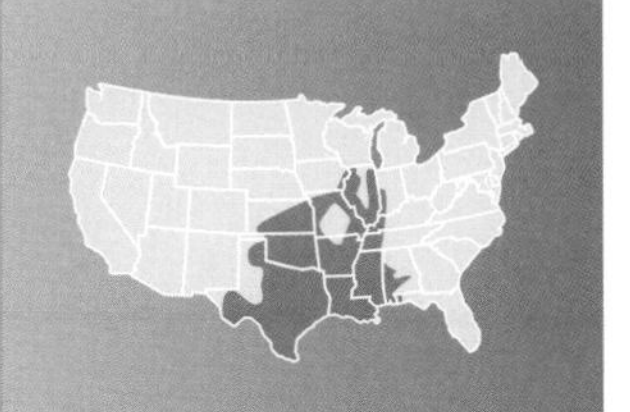

Baby diamondback watersnakes resemble the adults.

During cold or extremely hot periods they retreat to hiding spots in or near water such as beneath root masses, under logs, in the burrows of other animals, and beneath overhanging banks of rivers or lakes. They are primarily active in the water at night and can be found basking on limbs and other aquatic vegetation during the day.

FOOD AND FEEDING Diamondback watersnakes eat a wide variety of fish, including catfish, eels, sunfish, mullet, and mosquitofish, as well as frogs and toads. Adults eat mostly catfish. They sometimes feed by swiping the open mouth through the water and snapping at anything they happen to touch. They also exhibit the same behavior as brown watersnakes, wrapping the tail around a limb and dangling the body into the water to catch passing fish. These snakes are sometimes observed in high densities feeding on discarded fish carcasses near marinas.

REPRODUCTION Diamondback watersnakes give birth to live young. Large females sometimes produce litters of more than 50 young, but the typical litter size is about 25. In the spring, males apparently find females by scent trailing, and mating may occur on land or in water, often with more than one male attempting to mate with a single female. The babies are born in the late summer or fall.

PREDATORS AND DEFENSE Diamondback watersnakes commonly fall prey to a variety of natural predators typically found in lakes, rivers, and reservoirs of the Southeast, including cottonmouths, large catfish, alligators, and wading birds. They are noted for biting savagely and discharging a nauseating musk when captured. Given the chance, however, they will always try to escape first.

CONSERVATION Diamondback watersnakes are common in many aquatic habitats. Although harmless to humans, their superficial similarity to cottonmouths and their co-occurrence with that species in many parts of their geographic range result in large numbers of diamondback watersnakes being killed by people in some areas.

Diamondback watersnake from Ouachita Parish, Louisiana

The body pattern and color of the brown watersnake does not vary appreciably anywhere throughout its geographic range.

Brown Watersnake *Nerodia taxispilota*

DESCRIPTION This heavy-bodied watersnake has a light brown back with dark brown blotches down the middle alternating with, but usually not connecting to, blotches on the sides. The yellow or cream belly has irregularly spaced dark spots. The head is distinct from the body and is triangular.

WHAT DO THE BABIES LOOK LIKE? The babies resemble the adults.

DISTRIBUTION AND HABITAT Brown watersnakes are found from southeastern Virginia through most of the Coastal Plain of the Carolinas south through Florida and west to southern Alabama. They appear to be less ubiquitous in aquatic habitats than their western counterpart, the diamondback watersnake. They live in large rivers and adjacent swamps, small streams, and reservoirs created by damming rivers or streams. They are not as prevalent in other wetland habitats within their geographic range and are seldom found far from the water.

BEHAVIOR AND ACTIVITY Brown watersnakes hibernate in the northern parts of their range but may remain active during the winter in southern regions. Along waterways where they are common, they may be seen during the day in warm weather basking on rocks or on the limbs of trees and bushes, usually directly above the water, into which they retreat when disturbed. They actively forage at night and during the day.

How do you identify a brown watersnake?

SCALES
Keeled

ANAL PLATE
Divided

BODY SHAPE
Heavy bodied; broad, triangular head

BODY PATTERN AND COLOR
Dark brown, with square blotches on the back alternating with smaller blotches on the sides

DISTINCTIVE CHARACTERS
Eyes set more on top of the head than on the sides like most other snakes

SIZE

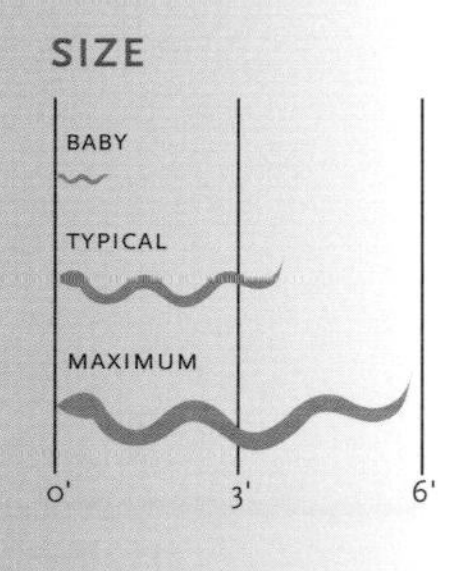

Brown watersnakes commonly bask in trees along streams and rivers.

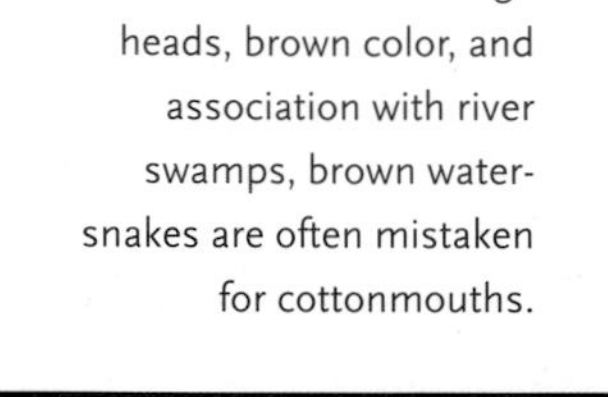

Because of their large heads, brown color, and association with river swamps, brown watersnakes are often mistaken for cottonmouths.

FOOD AND FEEDING Brown watersnakes feed almost exclusively on fish, especially catfish, which made up well over half of the reported food items in several studies of this species. Like diamondback watersnakes, brown watersnakes use an unusual ambush foraging tactic in which the snake coils its tail around a branch in the water and waits for fish to come within striking range. At other times they actively swim around, probing underwater holes and crevices in search of prey. Occasionally, a brown watersnake is found with the lateral spine of an ingested catfish protruding from its side. The spine eventually decomposes, and the snake apparently recovers with little difficulty.

REPRODUCTION Brown watersnakes mate in the spring, and one to three males may accompany a single female and attempt to mate with her. Males apparently find females by following their scent trails. Mating snakes have been observed on the ground and in trees overhanging water. Large females give birth to up to 61 babies, but the typical litter size is about 20–30. Babies are usually born on land in the late summer or fall.

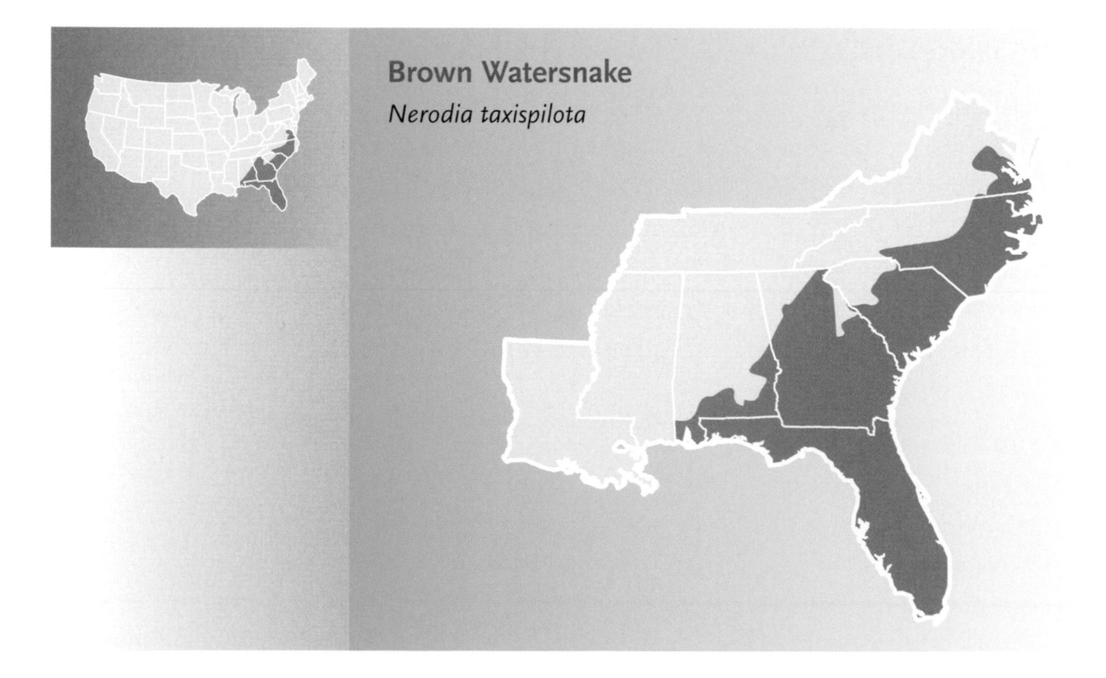

PREDATORS AND DEFENSE Because of their frequent association with large streams and rivers, brown watersnakes are eaten by a wide variety of natural predators, including hawks, large fish, cottonmouths, and alligators. They usually retreat quickly to the water when disturbed, but if captured will bite and release musk.

CONSERVATION Like the other large watersnakes that occur within or near the geographic range of the cottonmouth, brown watersnakes are occasionally killed in the mistaken belief that they are venomous. Along some rivers in the Southeast, brown watersnakes can be extremely abundant.

Did you know?

Watersnakes that feed on catfish are frequently found with the fish's spines sticking out through their body walls. The snakes generally recover without complications.

Eastern green water-snake from Hernando County, Florida

How do you identify an eastern green watersnake?

SCALES
Keeled

ANAL PLATE
Divided

BODY SHAPE
Heavy bodied

BODY PATTERN AND COLOR
Gray or greenish with yellowish or cream-colored belly

DISTINCTIVE CHARACTERS
Row of small scales between eye and upper lip scales

SIZE

BABY

TYPICAL

MAXIMUM

0' 3' 6'

Eastern Green Watersnake *Nerodia floridana*

DESCRIPTION This heavy-bodied watersnake is solid grayish or greenish above with a yellow belly that becomes darker under the tail. A row of scales encircles the lower half of the eye, separating it from the upper lip scales; among southeastern snakes, only the western green watersnake has a similar arrangement.

WHAT DO THE BABIES LOOK LIKE? The babies are similar to adults but with faint crossbands.

OTHER NAMES This species is called the Florida green watersnake by some herpetologists, although it is also found in Georgia and South Carolina.

DISTRIBUTION AND HABITAT Eastern green watersnakes are found throughout most of Florida, sporadically in southern Georgia, and in parts of the southern half of South Carolina in open, marshy wetlands with minimal tree cover. They are not typically residents of rivers, streams, or swamps, but sometimes occupy reservoirs.

BEHAVIOR AND ACTIVITY Like other large watersnakes of the Southeast, eastern green watersnakes hibernate in the northern, colder parts of their range but are active year-round in southern Florida. Even in areas with cold temperatures, they bask on sunny days in late winter, and if conditions remain warm they may stay active throughout the spring, summer, and fall. They frequently travel overland, especially in southern Florida during summer rains.

FOOD AND FEEDING Little is known about the diet of eastern green watersnakes. Most food records are of fish (e.g., sunfish, bass, and crappies) and frogs (e.g., pig frogs). Nothing is known of their hunting techniques, but they presumably employ an active foraging strategy. Like other watersnakes, they swallow their prey alive.

Eastern green watersnakes have a row of scales between the eye and lip scales.

REPRODUCTION The details of reproduction are poorly known. A few matings have been observed in the late winter or early spring. Females give birth in the summer, sometimes to huge litters. The record litter is 132 babies, which were dissected from a dead female, but the typical litter size is much smaller, about 20–40 young.

PREDATORS AND DEFENSE As is the case with other animal species that have large numbers of offspring, most baby eastern green watersnakes die before reaching adulthood. Predators characteristic of the wetland habitats frequented by this species include river otters, wading birds, hawks, ospreys, turtles, kingsnakes, alligators, and predatory fish. If unable to escape, eastern green watersnakes usually bite and release musk from scent glands in the cloaca.

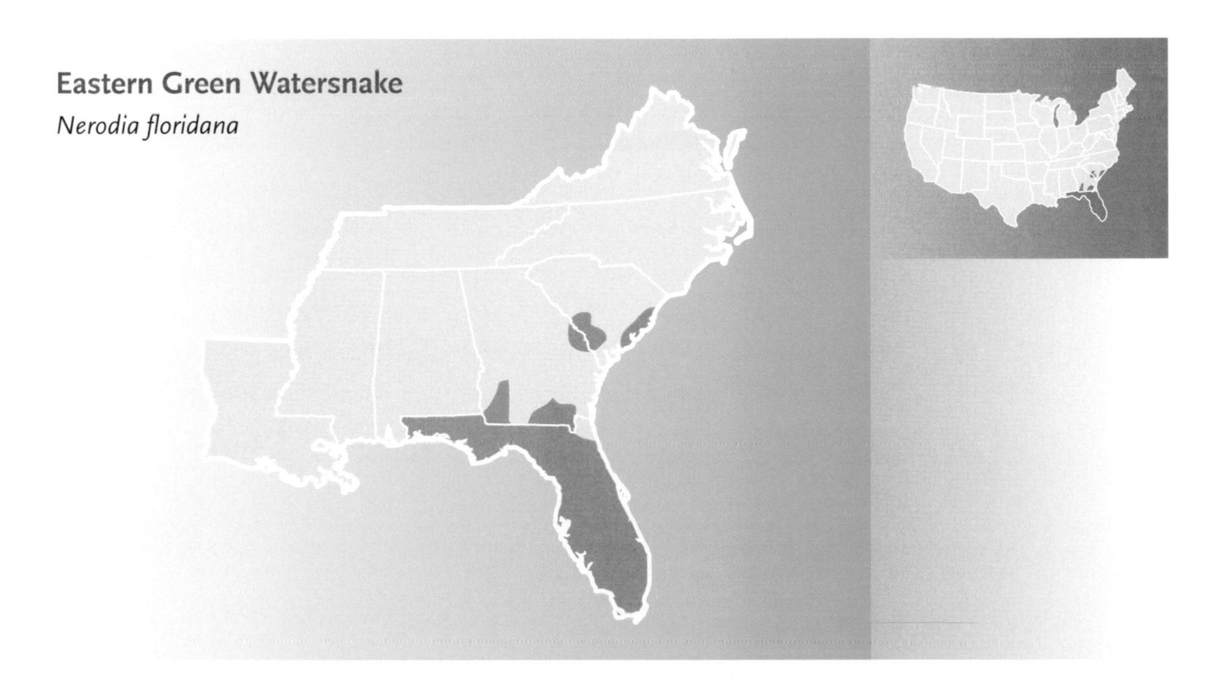

Eastern green watersnakes bask on cool days.

Some eastern green watersnakes from central and southern Florida are reddish in color.

CONSERVATION Eastern green watersnakes are very abundant in some aquatic ecosystems; however, their distribution in the northern parts of their range is patchy, and more information is necessary to determine their true status there. Occasionally, people kill eastern green watersnakes intentionally, and thousands die annually on Florida highways that are adjacent to wetlands.

Western green watersnakes often bask along waterways.

Western Green Watersnake *Nerodia cyclopion*

DESCRIPTION Western green watersnakes are generally solid grayish, greenish, or olive above, sometimes with faint dark bands, and have a dark belly with irregularly spaced yellowish spots. Like the eastern green watersnake, this species also has a row of scales encircling the lower half of the eye and separating it from the upper lip scales.

WHAT DO THE BABIES LOOK LIKE? Babies are similar to adults but with more distinctive crossbands.

OTHER NAMES Some herpetologists refer to this species as the Mississippi green watersnake, although it occupies several other states within and outside the Southeast.

DISTRIBUTION AND HABITAT Western green watersnakes occur in the Southeast along the Gulf Coast from the western tip of the Florida panhandle into Texas and up the Mississippi River drainage to Tennessee. They are common residents of swamps, bayous and ditches, open marshes, and small lakes and ponds.

BEHAVIOR AND ACTIVITY Western green watersnakes become inactive during the winter wherever and whenever cold temperatures occur. They generally hibernate under vegetation, in holes along the bank, or in stumps

How do you identify a western green watersnake?

SCALES
Keeled

ANAL PLATE
Divided

BODY SHAPE
Heavy bodied

BODY PATTERN AND COLOR
Gray or greenish with mottled gray belly

DISTINCTIVE CHARACTERS
Row of small scales between eye and upper lip scales

SIZE

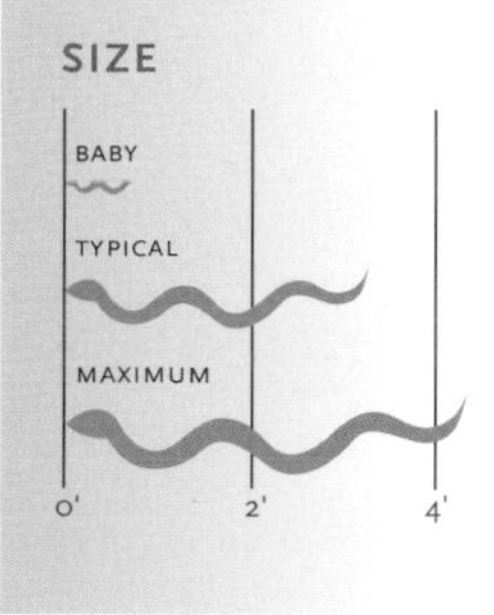

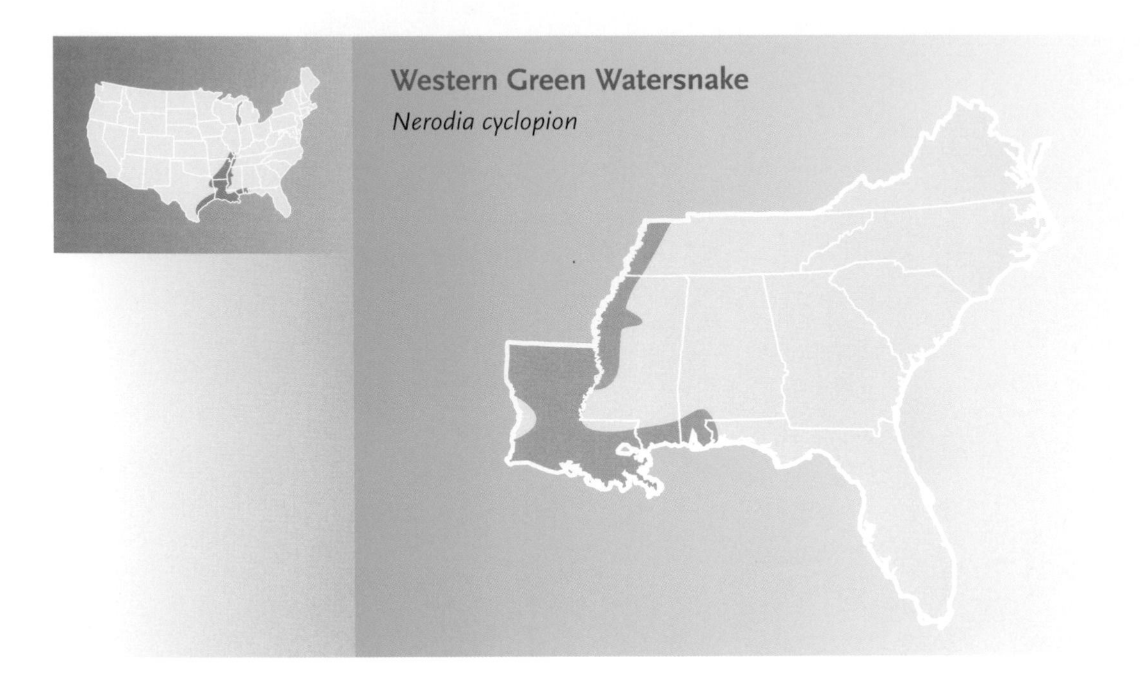

or rotten logs close to wetland habitats, and seldom venture far from water. The species is active in the daytime during the spring and fall and is also active at night during the summer months. Individuals frequently can be seen basking on logs and limbs above water. Western green watersnakes commonly travel overland between wetlands and are often found crossing roads.

FOOD AND FEEDING These apparently opportunistic aquatic foragers feed on many species of fish and amphibians. Little is known of their foraging strategy, but presumably they actively search out food using scent as their main method of prey detection. Prey items are seized and swallowed alive.

Western green watersnakes eat fish and amphibians.

REPRODUCTION Little is known about reproduction in this species. Most of the evidence points to mating in the spring, and many males may try to court a single receptive female. Females presumably announce their readiness to mate by releasing pheromones detectable by males. Young are born in the summer, and litters range in size from 6 to 37 (average = about 17).

Western green watersnakes have a row of scales between the eye and lip scales.

PREDATORS AND DEFENSE Western green watersnakes can fall victim to any of the typical predators inhabiting southern swamps, such as raccoons and river otters, wading birds, cottonmouths, kingsnakes, and alligators. They bite viciously when captured by another animal or a person, and like other watersnakes release a foul-smelling musk on the attacker.

CONSERVATION Besides habitat destruction, the greatest threats to western green watersnakes are vehicles on highways passing through swampy areas and intentional killing because of negative attitudes about snakes.

The posterior portion of the belly of the western green watersnake is dark.

Mud snakes usually have a shiny appearance.

How do you identify a mud snake?

SCALES
Smooth

ANAL PLATE
Usually divided

BODY SHAPE
Robust, with rounded head

BODY PATTERN AND COLOR
Shiny black above; belly with black-and-red checkerboard pattern

DISTINCTIVE CHARACTERS
Sharp spine on end of tail

SIZE

BABY
TYPICAL
MAXIMUM
0' 3' 6'

Mud Snake

Farancia abacura

Mud snakes have a sharp spine on the end of the tail.

DESCRIPTION Mud snakes are heavy-bodied, shiny black snakes with a red-and-black checkerboard pattern on their bellies. The red belly coloration extends onto the sides of the body in some individuals. In some cases, particularly on younger snakes, the red coloration may extend far up onto the sides of the body, appearing as thin bands. The rounded head is not very distinct from the neck, and there is a sharp spine on the tail tip. A few individuals have white pigment instead of red. Two subspecies, the eastern mud snake (*F. a. abacura*) and the western mud snake (*F. a. reinwardtii*), have been described and intergrade over a broad geographic area (see map).

WHAT DO THE BABIES LOOK LIKE? Baby mud snakes look like the adults, but the red markings sometimes extend farther up onto the body, sometimes forming complete red bands on the neck.

Mud Snake
Farancia abacura

OTHER NAMES Mud snakes are known as hoop snakes in some parts of the South because of their habit of lying in a loose coil, or horn snakes because of the tail spine.

DISTRIBUTION AND HABITAT Mud snakes range throughout the southeastern Coastal Plain from southern Virginia through all of Mississippi and Louisiana and western Tennessee. Adults occupy a variety of aquatic habitats, including cypress swamps, ponds, oxbow lakes, river floodplains, and even isolated wetlands—wherever their primary prey, amphiumas (large aquatic salamanders), are present. Young mud snakes are found in aquatic habitats near the area where they hatched, including smaller wetlands where the large salamanders may be absent but tadpoles and salamander larvae are abundant.

The red markings often extend farther up the sides on baby mud snakes.

BEHAVIOR AND ACTIVITY Mud snakes take refuge from cold weather by crawling into burrows made by other animals as well as under leaf litter, rotten logs, dead palmetto fronds, and other ground debris in terrestrial habitats near aquatic areas. They may be active

Rarely, mud snakes lack red pigment entirely.

The red-and-black belly is characteristic of mud snakes.

at any time of the year, including warm periods in the winter. Mud snakes are active on land and in the water both at night and during the day throughout much of the warm part of the year. Recently hatched young are commonly found on land near wetlands in the early spring.

Giant salamanders (genus *Amphiuma*) are the primary prey of adult mud snakes.

FOOD AND FEEDING Mud snakes specialize on amphiumas ("congo eels") but will also eat sirens, both of which are large, aquatic salamanders; juveniles prey on various species of aquatic salamanders, frogs, tadpoles, and occasionally fish. The enlarged teeth in the rear of the mouth and a sharp spine on the end of the tail may help the snake hold on to slippery prey.

REPRODUCTION Mud snakes apparently mate in the spring or early summer. The females lay their eggs in rotting vegetation, under logs, or in similar places during late spring or summer. Clutches contain from 4 to 104 eggs, with an average of about 30. Females often curl around their

When threatened, some mud snakes curl their tail in a defensive display.

Did you know?

The mud snake is one of more than 20 southeastern snakes that typically do not bite people, even when handled.

eggs and stay with them until they hatch in late summer or early fall. Baby mud snakes do not enter the water upon hatching but spend their first winter on land.

PREDATORS AND DEFENSE Predators presumably include terrestrial carnivores such as raccoons, racers, and birds of prey, as well as aquatic ones such as large fish, cottonmouths, and alligators. When threatened, some mud snakes display the red undersurface of the tail in a spiral, possibly to divert attention away from the head. Captured mud snakes will often coil around the captor's arm and harmlessly poke the tail spine against the person's skin. Mud snakes do not bite people, but they may release a foul-smelling musk.

CONSERVATION Among the greatest threats to mud snakes are the destruction and degradation of wetland habitats and the areas where these interface with the upland nesting habitats.

Adult rainbow snakes have highly specialized diets consisting primarily of American eels.

How do you identify a rainbow snake?

SCALES
Smooth

ANAL PLATE
Usually divided

BODY SHAPE
Robust body, rounded head

BODY PATTERN AND COLOR
Black above with three thin red stripes; chin yellow; belly red with black dots

DISTINCTIVE CHARACTERS
Sharp spine on end of tail; red stripes on back

SIZE

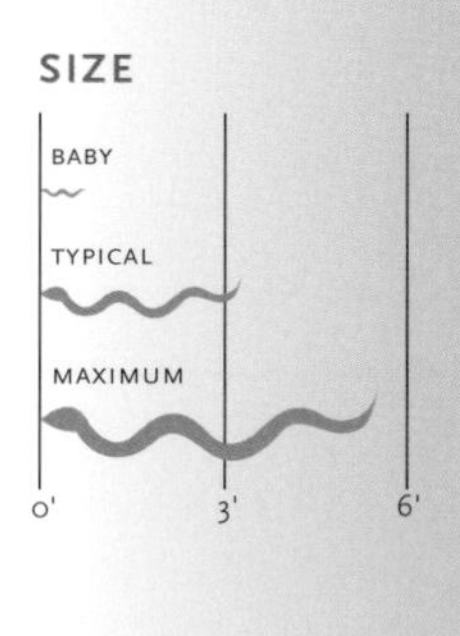

Rainbow Snake *Farancia erytrogramma*

DESCRIPTION Rainbow snakes are shiny black, heavy-bodied snakes with three thin red stripes running the length of the back. The belly is mostly red or pinkish with two or more rows of large black dots. The red often extends up onto the sides. The head is round and indistinct, and the tail has a sharp spine. Two subspecies, rainbow snake (*F. e. erytrogramma*) and South Florida rainbow snake (*F. e. seminola*), have been described.

WHAT DO THE BABIES LOOK LIKE? The young are identical to adults.

DISTRIBUTION AND HABITAT Rainbow snakes are restricted to the Southeast, being found throughout most of the Coastal Plain from Virginia to the southern half of Mississippi and into Louisiana. Adults are found in association with streams, river systems, swamps, and other aquatic habitats inhabited by their primary prey, American eels. They characteristically nest in upland areas

Baby rainbow snakes look similar to the adults.

Rainbow Snake

Farancia erytrogramma

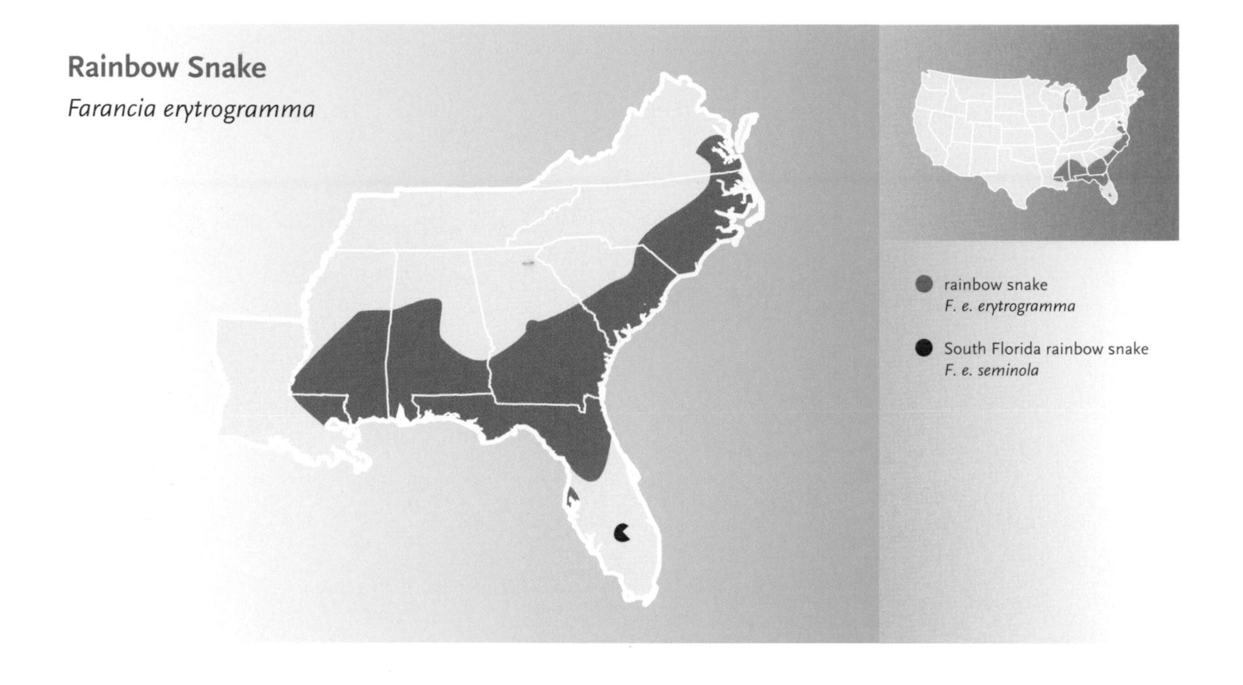

with sandy soil, and the young spend their first year or two in smaller isolated wetlands that have an abundant supply of tadpoles and salamander larvae.

Rainbow snakes have smooth scales and are usually shiny.

BEHAVIOR AND ACTIVITY Rainbow snakes are inactive during the winter, but both young and adults become active by mid-spring. Hibernation and warm-weather refuge sites are presumably beneath living root masses, in rotten stumps and logs, and in animal burrows or decayed root holes in and around river swamps. They are primarily active at night in the water and have been found crawling on land during both day and night hours. Rainbow snakes that are ready to shed are occasionally found out of the water basking on floating logs along stream or river banks.

FOOD AND FEEDING The principal prey of adult rainbow snakes is American eels; juveniles eat a variety of aquatic salamanders and tadpoles. As is the case with their close relatives the mud snakes, the tail spine of rainbow snakes may help them hold on to their slimy prey.

Rainbow snake from Aiken County, South Carolina

REPRODUCTION Little is known of the mating habits or other aspects of the reproductive biology of rainbow snakes. They are believed to mate in the spring; the eggs are laid in underground nests in sandy soil during the summer. Clutches range in size from 10 to 52 eggs. The eggs hatch during late summer or fall, but the young do not enter the aquatic environment until the following spring.

PREDATORS AND DEFENSE Like mud snakes, rainbow snakes probably fall victim to both terrestrial and aquatic predators. They are quite docile when captured by humans and do not bite.

CONSERVATION Habitat preservation can be critical to the long-term persistence of this species. Stream or river alterations and commercial overfishing that affects the abundance of American eels, their primary food, will have direct impacts on rainbow snakes. Human modifications, including highways, that create barriers or increase mortality of adult females traveling from lowland swamps to upland habitats to lay eggs could severely reduce rainbow snake populations in an area. The South Florida rainbow snake, only from Glades County, has not been found since the mid-1960s.

VENOMOUS SNAKES

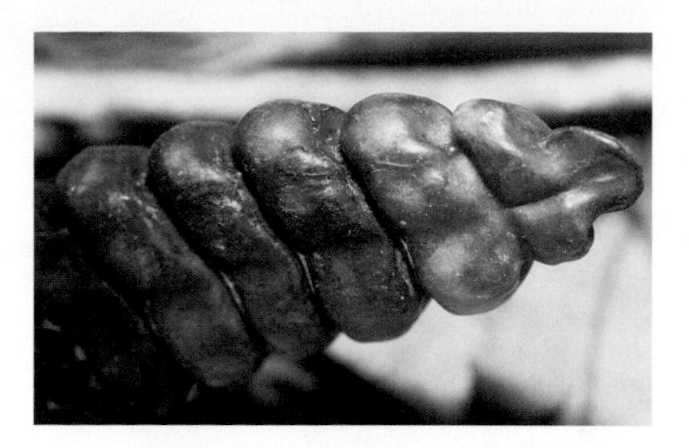

A rattlesnake rattle is composed of loosely connected segments of which one is added each time the snake sheds its skin.

Venomous snakes are among the most maligned and misunderstood animals on earth. Perhaps only sharks have generated more misconceptions and misinformation. To really understand venomous snakes, a person must view them in a practical and objective manner. Venomous snakes have fangs and venom for capturing prey, not to bite people. Because no venomous snake preys on humans, the only reason they bite humans is in self-defense, and then only as a last resort.

Cottonmouths have a reputation among the general public, and even among some scientists, for being extremely aggressive. Many people believe cottonmouths will actually pursue them with the intent of biting. In fact, a study done in South Carolina provided strong evidence that cottonmouths are very reluctant to defend themselves by biting. In that study, conducted in the swamps along the Savannah River, investigators tested how more than 50 cottonmouths responded to being stepped on and picked up by humans (*note*: the investigators used snake-proof boots to do the "stepping on" and an artificial arm to do the "picking up"). The cottonmouths usually first tried to escape. When stepped on, they usually tried to bluff by gaping widely, vibrating the tail, and releasing musk, but only rarely (less than 5% of the time) did they bite the boot. Even when picked up, only about 35% of the snakes bit the artificial arm, and most of those snakes were already aggravated from having been stepped on! The results of this study should help to dispel a common misconception about a species of venomous snake unjustifiably perceived as aggressive.

Only 6 of the 52 species of snakes found in the Southeast are venomous: the copperhead, the cottonmouth, the coral snake, and the three species of rattlesnakes. No single species occurs over the entire Southeast, and in most areas the number of venomous species is usually fewer than six. Additionally, the rarity of some species in many areas makes it extremely unlikely that a person will ever encounter them.

A young copperhead waits in ambush for a lizard or small frog.

Did you know?

Snakes do not chase people. If a snake moves toward a person, it probably just happens to be moving in that direction; or it may be trying to get to a hole or some other hiding place, and the person is in the way.

Pit vipers have elliptical pupils.

The skull of a rattlesnake showing the movable fangs and replacement fangs

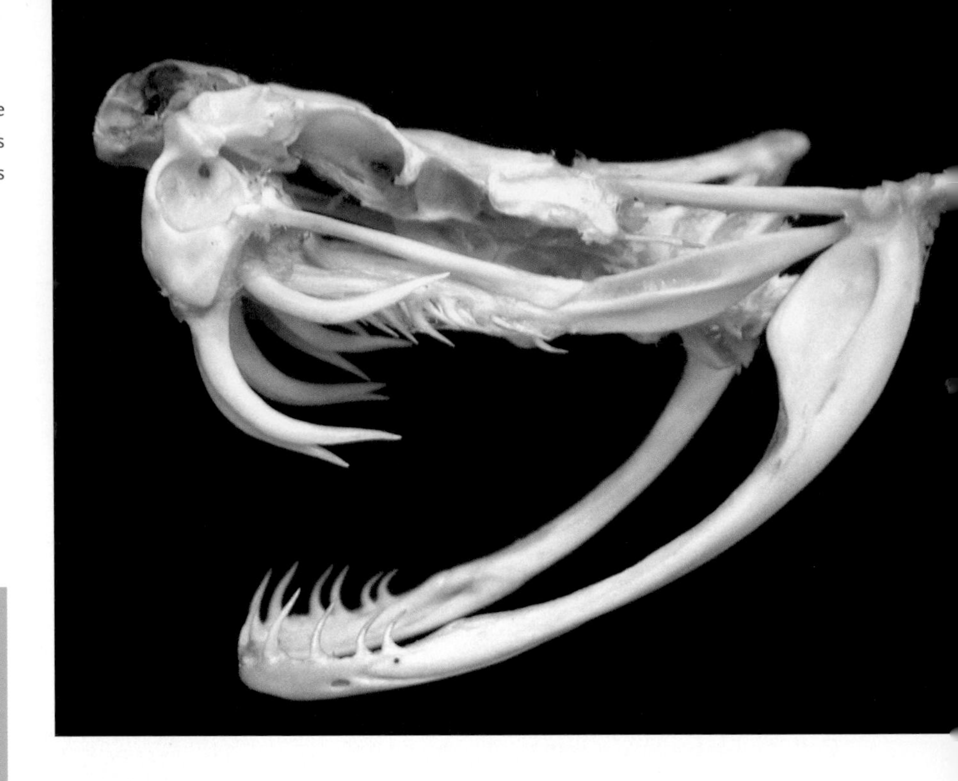

Did you know?

The venom of young pit vipers may be different from that of adults if young snakes feed on different prey than their parents.

VENOMOUS OR POISONOUS?

Many people refer to venomous species of snakes as "poisonous" snakes. In fact, the word *venomous* is more appropriate than *poisonous* because of the way the toxin is administered to the victim. Venom is a toxic substance produced in special glands by one animal that is injected into the body of another animal mechanically, such as by stinging or biting. A poison, in contrast, is a substance that is toxic when consumed or touched. Thus, southeastern pit vipers and the coral snake are better referred to as venomous snakes because they inject toxins into other animals through their hollow fangs.

BIOLOGY OF VENOMOUS SNAKES

The cottonmouth, copperhead, and rattlesnakes are all pit vipers (family: Viperidae, subfamily: Crotalinae). Vipers have movable fangs in the front of the mouth that fold up against the roof of the mouth when not in use. Pit vipers also have "pits," special heat-sensing organs that look like holes between the eye and nostril. The pits sense infrared radiation, such as heat produced by warm-blooded prey. All vipers in the Americas are pit vipers, including tropical species such as the bushmaster and fer-de-lance.

The venom of most pit vipers contains *hemotoxic* proteins that destroy blood and tissue, although the venom's toxicity can vary among snake species. Diamondback rattlesnakes, for example, have fairly potent hemotoxic venom, but the venom of copperheads is generally less so.

The coral snake is the only venomous snake native to the Southeast that is not a pit viper. The coral snake is a member of the family Elapidae, which includes the cobras, kraits, and mambas. Members of this family have short fangs in the front of the mouth that do not fold up like those of vipers, and they lack heat-sensitive pits to detect warm-blooded prey. Their venom is primarily *neurotoxic* and acts by inhibiting the nervous system—including the nerves that control breathing. Prey animals usually die because they are unable to breathe.

Did you know?

Among snakes that use venom to defend themselves, juveniles may be more likely to strike and bite, perhaps because their smaller size makes them feel more vulnerable. However, young snakes' venom is not necessarily more potent, and young venomous snakes cannot inject nearly as much venom as adults.

SNAKEBITE

One of the biggest misconceptions about snakes is the extent of the danger they pose to humans. In actuality, in the United States, a person is far more likely to be killed by lightning than to die from a venomous snake's bite. Certainly, some snakes can seriously harm or even kill humans. But although hundreds of people are bitten each year by copperheads (which are responsible for most of the venomous snakebites in the Southeast), only one human death has ever been documented, and no deaths from pigmy rattlesnakes or eastern coral snakes were reported between 1983 and 2003. According to the Centers for Disease Control, more people are seriously bitten by domestic dogs every year in this country (800,000) than by venomous snakes (7,000). Like many other wild animals, as well as domestic pets such as dogs and cats, venomous snakes can be dangerous. But snakes are also natural and important parts of the ecosystems of the Southeast.

Using common sense is the best way to minimize your chance of getting bitten by a venomous snake. Become familiar with the snakes in your area. Never try to capture or handle a venomous snake—or *any* snake of unknown identity. Children should be taught to appreciate snakes but to leave *all* of them alone unless they are with an adult who can distinguish between venomous species and harmless ones. When you are outside in areas where venomous snakes occur, use a flashlight at night, wear sturdy shoes, and watch where you step and put your hands.

The tiny rattle of a pigmy rattlesnake

Snakebite kits The best snakebite kit is a set of car keys, a cellular phone, and a companion.

SNAKEBITE TREATMENT

A bite from a venomous snake should be treated as a medical emergency. First, try to remain calm. Remember, the probability of dying from a venomous snakebite in the United States is extremely low. With proper medical treatment, most people recover fully and with few complications. Seek medical attention as quickly as possible. DO NOT try to cut the snake-bitten area. DO NOT use a tourniquet. DO NOT drink alcohol. DO NOT apply ice to the area or use electric shock treatment. Most snakebite kits do little, if anything, to lessen the severity of snakebite, and some, such as the old kits that promoted cutting and use of a tourniquet, are likely to do more harm than good.

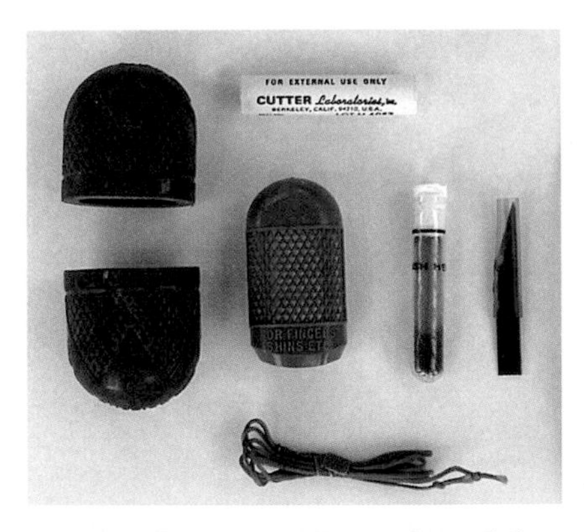

You should **not** use the old snakebite kits that require cutting and use of a tourniquet.

If a child is bitten by a venomous snake, the consequences could be more serious simply because the same quantity of venom will have proportionately more impact on someone with a smaller body size. The good news is that a study of more than 1,000 venomous snakebites in the United States found that in more than 60% of cases, the snake injected little or no venom. Consequently, what may appear to be a life-threatening situation may actually turn out to be medically insignificant. Nevertheless, the safest approach is to get anyone, child or adult, bitten by a venomous species to an emergency care facility as quickly as possible.

Southern copperhead from Liberty County, Florida

Copperhead

Agkistrodon contortrix

DESCRIPTION Copperheads have dark brown, hourglass-shaped crossbands on a light brown background. The crossbands may be broken in places. The belly is pale brown with irregularly spaced darker markings. The background color of the southern copperhead (*A. c. contortrix*) is rather light and sometimes looks pinkish. The northern copperhead (*A. c. mokasen*) has wider crossbands than the southern subspecies and is usually darker overall. The two subspecies intergrade over a very broad geographic area (see map).

WHAT DO THE BABIES LOOK LIKE? The babies have the same pattern as the adults but have yellow or green tail tips.

OTHER NAMES This species is called the highland moccasin in some rural areas.

DISTRIBUTION AND HABITAT Copperheads are both wide ranging and common. They occur throughout every southeastern state except Florida, where they barely enter the Panhandle, and the southeastern one-third of Georgia. They are found in all types of terrestrial habitats, including rocky areas in the mountains; mixed hardwood and pine forests; swamp margins; and even farms, suburban areas, and coastal island resorts if sufficient ground vegetation or other cover is available for concealment.

How do you identify a copperhead?

SCALES
Keeled

ANAL PLATE
Single

BODY SHAPE
Heavy bodied, but less so than other pit vipers

BODY PATTERN AND COLOR
Light brown body with darker, hourglass-shaped crossbands

DISTINCTIVE CHARACTERS
Top of head solid coppery brown

SIZE

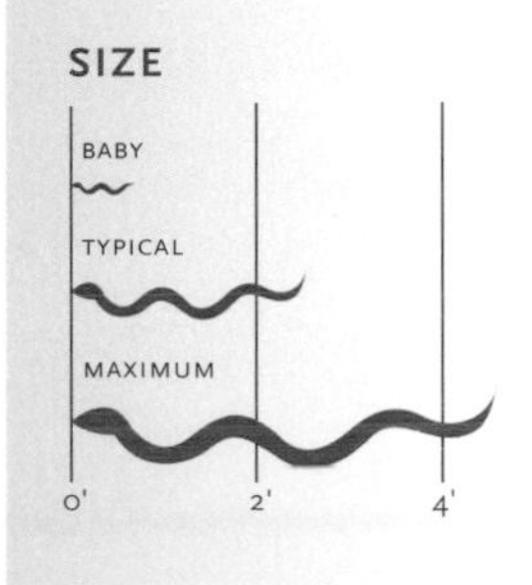

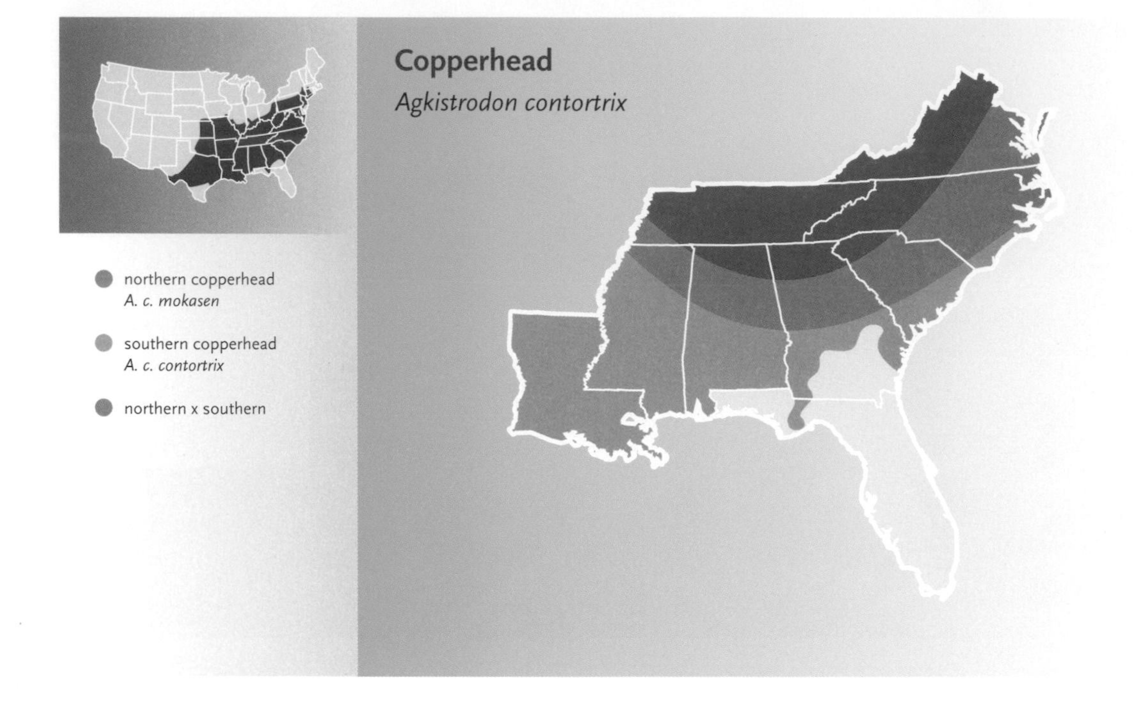

BEHAVIOR AND ACTIVITY Copperheads are noted for hibernating in communal dens with several to dozens of other individuals, but such behavior is most characteristic of northern and mountain populations. They sometimes hibernate communally with timber rattlesnakes. Southern Atlantic and Gulf Coastal Plain copperheads are more likely to hibernate alone, but occasionally several can be found in a suitable hibernation site. Stumps with old root holes, rocky crevices, and other underground sanctuaries are common retreats of copperheads. Copperheads spend many hours of the year aboveground, either coiled in a camouflaged setting such as leaf litter or moving about. They are generally more active at night but may be found in the daytime during cool periods.

Copperheads can detect warm-blooded prey at night by using their sensory pits.

Juvenile copperheads have bright yellow tails.

FOOD AND FEEDING Copperheads are opportunistic predators that eat a wider variety of prey than most other venomous snakes. The diet of adults consists mostly of vertebrates such as voles, mice, birds, frogs, lizards, and other snakes; they will even eat large insects. They usually wait in ambush for their prey but sometimes actively forage and have been observed climbing trees to capture cicadas. Juveniles enhance their ability to ambush prey by wiggling their yellow tail tip, which attracts small lizards and frogs within striking range. A copperhead feeding on a large prey animal such as a rodent strikes the prey and releases it, and then tracks the dying animal by following its odor trail. Other prey, such as frogs and birds, are struck and held until the venom takes effect.

REPRODUCTION Copperheads have two mating seasons: late spring and early fall. After mating, the females can store sperm in special receptacles for long periods before fertilization occurs. The male finds a mate by following her pheromone trail and then actively courts her by touching her with his snout and rubbing her neck with his chin. When a male copperhead encounters another male during the mating season, they may engage in a "combat dance" in which each tries to pin the other to the ground. The young are born in the late summer or early fall. Typical litter sizes are 7–8, but very large females may produce more than 20 babies. In the mountains, female copperheads congregate and give birth prior to entering communal hibernation dens. Most females give birth every 2 years, but they may reproduce annually if enough food is available.

How dangerous are they? More people are bitten by copperheads in the Southeast than by any other venomous snake. The bites can be very painful, but only one death has ever been documented. The venom of copperheads destroys red blood cells and other tissues but is not considered highly toxic to humans. However, if you are bitten, you should seek medical attention immediately. If you find a copperhead in the field, you can observe it from a safe distance but should not approach it. If a copperhead turns up in your backyard or some other area frequented by people, the best approach is to have the snake removed by someone experienced in handling venomous snakes.

PREDATORS AND DEFENSE Small copperheads presumably are preyed on by any animal that can avoid or will risk a mildly venomous bite. Adults are somewhat less vulnerable, but common kingsnakes, which are immune to the venom, will eat copperheads as readily as other prey. Copperheads respond to kingsnakes by lifting the center of the body upward to ward off attack rather than by attempting to strike. Copperheads' first line of defense is always camouflage or concealment, and they are masters at it. People unquestionably come within striking distance of innumerable copperheads each year but pass by unaware. Once discovered, a copperhead will strike if it feels threatened. A strike toward a person may be from several feet away with no chance of successfully biting, and is assumed to be a form of threat display. Copperheads release a musky odor when disturbed, and most will bite if they are picked up.

Northern copperhead from Watauga County, North Carolina

CONSERVATION Copperheads are not considered to be threatened by environmental degradation in most areas because they can live in a variety of habitats, but individual copperheads are persecuted by humans who fear being bitten by them. Unfortunately, many harmless snakes, such as rat snakes, corn snakes, and northern and banded watersnakes, are killed each year because of their superficial resemblance to copperheads.

Florida cottonmouths can get larger than the other subspecies.

Cottonmouth

Agkistrodon piscivorus

DESCRIPTION Cottonmouths have wide, dark crossbands on a brown to olive-brown background. Some large individuals may be almost solid black. The belly has irregularly spaced, dirty yellow and brown blotches, and the underside of the tail is usually black. The eastern cottonmouth (*A. p. piscivorus*) generally has a lighter background color and has no pattern on its snout. The western cottonmouth (*A. p. leucostoma*) looks like the eastern cottonmouth but is usually darker. The Florida cottonmouth (*A. p. conanti*) is usually very dark to solid black and has two dark vertical bars on its nose. Cottonmouths that live in water with high concentrations of tannins can develop a solid coppery color. A broad zone of intergradation occurs among subspecies of the cottonmouth (see map).

WHAT DO THE BABIES LOOK LIKE? Baby cottonmouths are lighter in color and more boldly patterned than adults and have a yellow to greenish tail tip. Baby cottonmouths are frequently misidentified as copperheads.

Baby cottonmouths are brightly banded and look similar to copperheads.

How do you identify a cottonmouth?

SCALES
Keeled

ANAL PLATE
Single

BODY SHAPE
Heavy bodied, some large individuals extremely so

BODY PATTERN AND COLOR
Black to olive-brown, sometimes with dark bands

DISTINCTIVE CHARACTERS
Black line from eye to back of lower jaw

SIZE

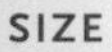

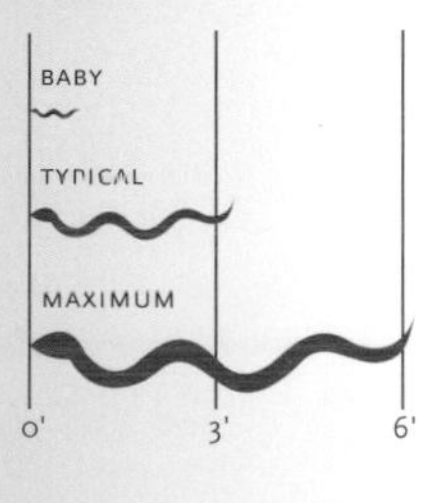

Eastern cottonmouth from Aiken County, South Carolina

OTHER NAMES Despite attempts by herpetologists, state game officials, and conservationists to declare that the name for this species is cottonmouth, the names water moccasin and cottonmouth moccasin are deeply ingrained throughout much of the South.

DISTRIBUTION AND HABITAT Cottonmouths range into at least part of every southeastern state but are absent from most of the Piedmont and mountainous areas. They are associated with river swamps, backwaters, and floodplains throughout their geographic range, but are less common on the rivers themselves. They are prevalent around lakes, reservoirs, small ponds, streams, and isolated wetlands, especially when such habitats are surrounded by swampy habitat or heavy vegetation.

BEHAVIOR AND ACTIVITY Cottonmouths hibernate during cold winter weather, but a few individuals can be found aboveground on sunny days, usually coiled near a hole into which they can readily escape. Hibernation retreats may be on land several hundred feet from water and are often in stump holes or holes left by decayed roots, under logs, or in abandoned animal burrows. Beaver lodges seem to be particularly attractive to cottonmouths in fall and winter, probably because of the many hiding places they offer both in and out of the water. In most of Florida, cottonmouths are active year-round.

During the active season, cottonmouths move around both in the daytime and at night, commonly hunting after dark during hot summer weather. They can be found at any time of day coiled on stumps, logs, or land associated with river swamps and other wetland habitats. On cool spring and fall days they climb into the lower branches of trees or bushes to bask. When traveling, cottonmouths are as likely to be seen on land as in the water, but they are seldom found far from aquatic habitats.

Did you know?

Snakes can bite underwater. Watersnakes and cottonmouths feed on fish, which they capture with their mouths underwater.

FOOD AND FEEDING Cottonmouths are opportunistic predators that eat a wide assortment of prey, which they capture using both ambush and active foraging methods. Some food is even acquired by scavenging. The diet includes small mammals, birds, fish, frogs, salamanders, turtles, other snakes (especially watersnakes), and even baby alligators. The cottonmouth usually strikes and holds its prey, though particularly dangerous animals such as rodents that might bite or scratch may be released and later tracked

How dangerous are they? If you find a cottonmouth in the field, leave it alone! Cottonmouths can be dangerous if harassed, but they can be safely observed from a distance of several feet. Take care when walking in known cottonmouth habitat. A snake may not move if it thinks it has not been seen, posing a potential hazard for someone who unwittingly steps on one. The venom of cottonmouths destroys blood cells and has anticoagulant properties, and deaths have resulted from their bites. However, nearly all victims who receive proper medical treatment survive.

by their scent. Juvenile cottonmouths attract prey the way young copperheads do, by wiggling their yellow tail tip.

Cottonmouths gape to reveal the white lining of the mouth during the open-mouthed threat display.

REPRODUCTION Cottonmouths mate primarily in the spring but may have a fall mating period as well. Males find mates by following pheromone trails. Like many other pit vipers, males "fight" with competing males in a "combat dance" to win the privilege of mating with available females. Female cottonmouths can store sperm for long periods. Litters of 1–20 babies (average = about 7) are born in late summer or early fall. Female cottonmouths sometimes congregate when giving birth. A female cottonmouth may have young at intervals of 2–3 years.

PREDATORS AND DEFENSE Among the most successful natural predators on cottonmouths are alligators, kingsnakes, and larger cottonmouths. Large mammals, wading birds, hawks, and owls are also known or suspected predators, but most of these species restrict themselves to juveniles because of the potentially greater venom hazard associated with adults. Cottonmouths rely on camouflage and concealment for defense and will usually remain motionless unless they are certain they have been discovered. The first response of many individuals to a perceived threat is to escape down a hole, under a stream bank, or into the water if time allows. If a safe and successful escape seems uncertain, most cottonmouths will stand their ground. The common name of the species is derived from the open-mouthed threat display that is characteristic of many, but not all, individuals. The lining of the mouth is white rather than the darker pink, red, or even black typical of other species of snakes. Other threat displays or defensive behaviors include rapidly vibrating the tail, flattening the head and body to look larger, and releasing a strong-smelling musk. Cottonmouths sometimes release musk before they are actually approached and can be located by the smell.

Behavioral research with cottonmouths in the field and laboratory has demonstrated conclusively that the species is not aggressive toward humans; their behavioral responses are solely defensive in nature. These studies have also shown that most cottonmouths will not bite a person unless they are picked up. Nevertheless, if you are walking in an area where cottonmouths occur, be careful where you step.

CONSERVATION Cottonmouths are not threatened as a species, but localized populations may be extirpated when wetland habitats are destroyed or highways are built between the aquatic habitats where cottonmouths feed and the upland sites where they hibernate.

Adult western cottonmouths are usually darker than eastern cottonmouths.

Canebrake rattlesnake from Surry County, North Carolina. The stripe from the eye to the corner of the jaw is absent in timber rattlesnakes.

How do you identify a timber or canebrake rattlesnake?

SCALES
Keeled

ANAL PLATE
Single

BODY SHAPE
Heavy bodied

BODY PATTERN AND COLOR
Gray, brown to almost black, or yellow with dark chevrons

DISTINCTIVE CHARACTERS
Large rattles and black tail

SIZE

BABY
TYPICAL
MAXIMUM
0' 3' 6'

Timber/Canebrake Rattlesnake *Crotalus horridus*

DESCRIPTION Timber and canebrake rattlesnakes, which are the same species, typically have a gray, brown, or yellow back with black chevrons or crossbands. The belly is yellow to grayish white stippled with black dots. Timber rattlesnakes usually have a yellowish brown or gray background but are sometimes completely black. Canebrake rattlesnakes are lighter in color, and the background coloration often has a pinkish hue. The canebrake also has an orange or brown stripe running the length of the back. Both forms have a solid black tail.

WHAT DO THE BABIES LOOK LIKE? The babies are replicas of adults but often are more strikingly marked.

DISTRIBUTION AND HABITAT Throughout most of their geographic range, especially in mountainous areas, these big

Young canebrake rattlesnakes look similar to the adults.

Timber/Canebrake Rattlesnake

Crotalus horridus

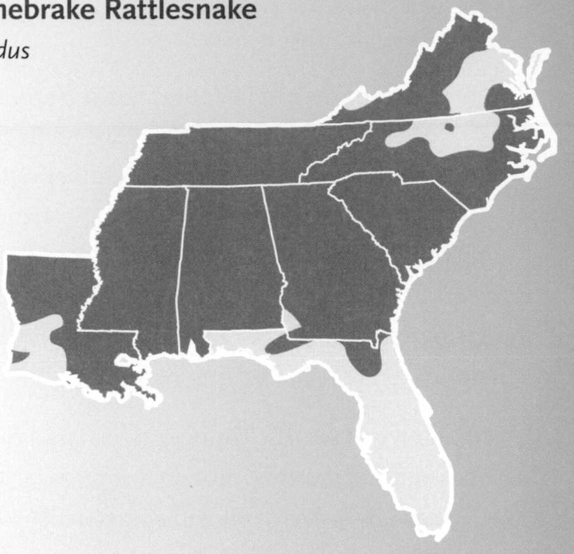

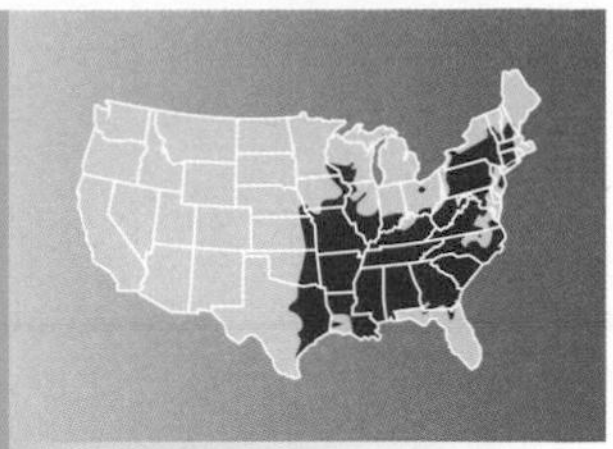

rattlesnakes are called timber rattlesnakes. Those found in the Coastal Plain are often called canebrake rattlesnakes. The species is found in every southeastern state but is absent from most of the Florida panhandle and peninsula, and from portions of North Carolina, Virginia, and Louisiana. Timber rattlesnakes occupy virtually all terrestrial habitats within their geographic range, including hardwood forests, pine flatwoods, rocky mountainous terrain, floodplains, high ground in swamps, and rural agricultural areas. Their persistence, however, is usually incompatible with human occupation, and few remain in urbanized or residential areas, or even in farming communities.

Did you know?

Among snakes of the Southeast, several species live more than 20 years, including rat snakes, indigo snakes, and copperheads. One timber rattlesnake lived more than 30 years in captivity.

BEHAVIOR AND ACTIVITY Timber rattlesnakes spend a few weeks to a few months in hibernation each winter, depending on the climate of the region and the winter weather of the year. Those in mountainous areas are noted for hibernating in communal dens with others of their species and with copperheads. Canebrake rattlesnakes in most of the Coastal Plain are more solitary, although an ideal den site such as a large stump hole with deep root tunnels or a rocky area with many crevices may attract several individuals of both sexes and all sizes. They also seek refuge in culverts and under pieces of tin and wood around old homesites.

These rattlesnakes generally become active in mid-to-late spring and may remain active into late fall. Both timber rattlesnakes and canebrake

The stripe down the center of the back is characteristic of the canebrake rattlesnake, the coastal plain counterpart of the timber rattlesnake.

rattlesnakes can be found moving around during the day, and the latter are also commonly found moving at night during hot weather. Timber rattlesnakes may travel 2–4 miles from den sites to foraging habitats or in search of mates. Less is known about canebrake rattlesnakes, but presumably they also move long distances during the spring and fall. The longest movements are made by large males during the fall mating season.

FOOD AND FEEDING Timber rattlesnakes eat small mammals such as shrews, mice, rats, squirrels, and rabbits almost exclusively, but will occasionally consume birds. Like other rattlesnakes, they are ambush hunters that sit motionless for extended periods waiting for prey to approach. Ambush points are usually along a log or in other places frequented by small mammals. Individuals sometimes sit at the base of a tree, head pointed upward, waiting for a squirrel to come down the trunk. They usually strike the prey, inject venom, and then release it, following the dying animal's scent until it can be safely eaten.

REPRODUCTION Female timber rattlesnakes may take 5–10 years to mature and are capable of reproducing every 2–3 years thereafter, although some may wait as long as 6 years in times of low food availability (e.g., during

How dangerous are they? Timber and canebrake rattlesnakes are potentially very dangerous and should be treated with the utmost respect. Although they are generally very reluctant to bite people, when bites do occur they can be very serious, and some have resulted in death. The venom destroys blood and tissues; some populations are reported to have venom with neurotoxic elements that affect the nervous system. If you come across one in the field, you can watch it from a safe distance, but never attempt to harass or capture the snake. If you find a timber rattlesnake or canebrake rattlesnake in areas frequented by people, arrange to have it removed by someone with experience handling venomous snakes.

a drought). Timber rattlesnakes mate primarily in the fall, though some breeding may occur in the spring. After mating, females store sperm in their oviducts over the winter and use it to fertilize their eggs in the spring. Litters of 6–18 are produced in late summer (canebrake rattlers) or early fall (timber rattlers). In the mountains, female timber rattlesnakes congregate to give birth in rookeries that provide good opportunities for thermoregulation during the final stages of gestation. Females usually stay with their babies for up to 2 weeks (until the first time they shed their skin).

PREDATORS AND DEFENSE Adults have few natural enemies because their venom protects them. Kingsnakes, indigo snakes, and opossums are immune to the venom of rattlesnakes, however, and probably attempt to eat those they encounter. Kingsnakes can kill and eat rattlesnakes that are almost the same length as they are. Carnivorous mammals and predatory birds probably prey on young canebrakes. Rattlesnakes of all sizes are susceptible to mammalian predators during the winter when their lower body temperatures leave them incapable of defending themselves effectively.

Did you know?

Female timber rattlesnakes may take 10 years to reach maturity and may reproduce only every 3–4 years.

The body color of timber rattlesnakes ranges from yellowish to nearly solid black.

Timber and canebrake rattlesnakes are relatively benign when encountered by humans. Their first response is usually to become motionless, remaining coiled, or stretched out if they were moving. Many will not even rattle. If pestered, they will rattle, coil, and assume a defensive pose, and most will bite if approached too closely or picked up.

CONSERVATION Rattlesnakes throughout the Southeast are threatened by habitat destruction and development, especially by the construction of highways that they must cross to reach seasonal habitats. In North Carolina they have been extirpated from most of the Piedmont as a result of human activity, and 84% of 200 canebrake rattlesnakes examined in a South Carolina study were found on highways, usually dead.

Did you know?

Snakes do not travel in pairs or groups. Snakes are usually found alone, although during the fall or spring breeding seasons, pairs of snakes can sometimes be found together. Some snakes, such as timber rattlesnakes, congregate at rocky outcrops to hibernate during the winter.

Timber rattlesnakes usually sit and wait for their prey.

SCIENTIFIC NOMENCLATURE Timber and canebrake rattlesnakes were once considered separate subspecies (timber rattler, *C. h. horridus*; canebrake rattler, *C. h. atricaudatus*). Canebrake rattlesnakes found in the Coastal Plain are very distinct in color pattern and habits from mountain-dwelling timber rattlesnakes, although populations in the Piedmont and other areas between lowlands and mountains often have individuals with intermediate characteristics.

Eastern diamondback rattlesnakes vary little in body color and pattern throughout their geographic range.

How do you identify an eastern diamondback rattlesnake?

SCALES
Keeled

ANAL PLATE
Single

BODY SHAPE
Heavy bodied

BODY PATTERN AND COLOR
Brown, gray, or yellow background with black, light-bordered diamonds

DISTINCTIVE CHARACTERS
Two light stripes running from eye to jaw

SIZE

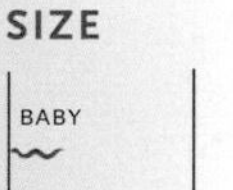

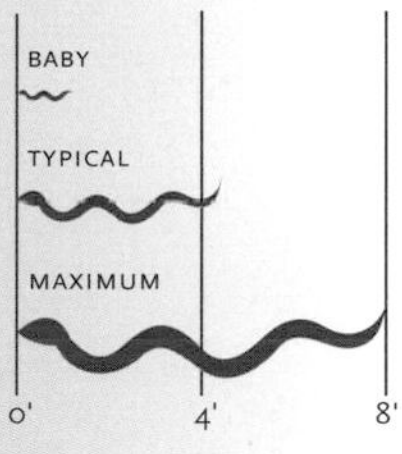

Eastern Diamondback Rattlesnake

Crotalus adamanteus

DESCRIPTION Eastern diamondback rattlesnakes are very heavy bodied snakes with large, dark, diamond-shaped markings outlined with white on a brown, gray, or yellowish background. The diamond markings often fade toward the tail. The unmarked belly is pale gray to white. Eastern diamondbacks have large rattles and two light stripes on either side of the head that start in front of and behind the eye and extend to the corner of the jaw.

WHAT DO THE BABIES LOOK LIKE? Baby eastern diamondbacks are miniature versions of the adults.

DISTRIBUTION AND HABITAT Eastern diamondbacks have the most limited geographic range of the southeastern pit vipers, historically inhabiting much of the Lower Coastal Plain from North Carolina to Louisiana in pine flatwoods, sandhill habitats, low-lying palmetto stands, and coastal islands with comparable habitat. They are often found in association with pine plantations managed for quail, presumably because they feed on the rodent populations that thrive around feeding stations, and in areas with abundant gopher tortoise burrows in which they seek shelter.

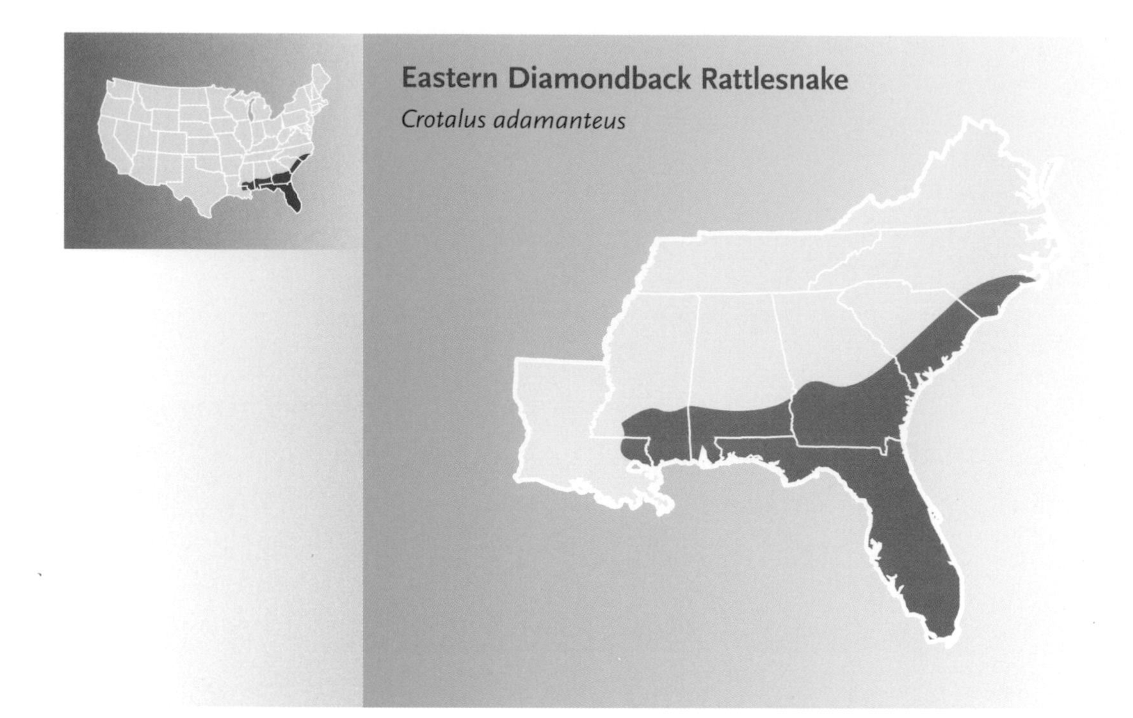

Eastern Diamondback Rattlesnake

Crotalus adamanteus

BEHAVIOR AND ACTIVITY Diamondback rattlesnakes become inactive during cold winter weather but do not ordinarily hibernate for long periods because prolonged cold periods are uncommon in their range. Stump holes with extensive networks of decayed root tunnels are preferred retreats from extremely hot or cold temperatures. In habitats where gopher tortoises are found, diamondback rattlesnakes often retreat deep into tortoise burrows. During the warmer months, diamondbacks spend much of their time coiled beneath palmetto bushes, grass clumps, and other vegetation, where their camouflage renders them almost invisible. Overland travel is almost always during daylight hours. Individual diamondbacks may move distances of a mile or more on some occasions, but they generally have smaller home ranges and travel shorter distances than timber rattlesnakes.

Eastern diamondback rattlesnake swallowing a cottontail rabbit

FOOD AND FEEDING Adults eat mammals such as cotton rats, squirrels, and rabbits. Young snakes feed primarily on smaller mammals such as voles and mice. Diamondbacks are ambush predators that may remain for weeks in one spot waiting for prey to come by. After striking and injecting venom, they release the animal and allow it to crawl away to die, then use their excellent odor detection abilities to locate it. Large adult diamondbacks, which feed primarily on large prey such as cottontail rabbits, may consume only four or five meals each year.

REPRODUCTION Eastern diamondback rattlesnakes mate mostly in the late summer or early fall, although spring mating has been reported. Mated females can store viable sperm for long periods. Baby diamondbacks are born during the early fall. The litter size is typically about 12 but can be more than twice that. Females usually give birth inside a tortoise burrow or old stump hole, and the mother stays with the babies, apparently to protect them, for a few weeks until they shed their skin for the first time. Eastern diamondback females reproduce only every 2–3 years.

PREDATORS AND DEFENSE The largest diamondback rattlesnakes have few natural enemies, although big kingsnakes and indigo snakes may eat some adults. Young diamondbacks probably fall prey to hawks and possibly some large mammals such as raccoons, skunks, and foxes, although the cost of being bitten by even a small diamondback may be a deterrent. Diamondback rattlesnakes coiled in a camouflaged position often will not reveal their presence, but those physically disturbed or confronted by an adversary while crawling on the surface are unquestionably among the most dangerous snakes in America. A diamondback on the defensive typically lifts its head several inches above the ground, keeping the neck in an S-shaped curve ready to strike while the tail rattles a warning. Once aggravated, the snake continues to rattle, sometimes backing away into underbrush to escape. Although their actions are strictly a form of defensive behavior and not of aggression, diamondbacks can be provoked unwittingly and may bite

Did you know?

You cannot age a rattlesnake by the number of its rattles. Rattlesnakes add a rattle segment every time they shed their skin, which may be several times each year. Also, rattle segments break off the end of the rattle string from time to time.

How dangerous are they? Eastern diamondback rattlesnakes are potentially the most dangerous snakes in the United States. Their venom is highly toxic, and adults can inject much more of it than smaller venomous species can inject. The venom rapidly destroys tissues and blood cells, and bites not treated with antivenom have resulted in deaths within 24 hours. Fortunately, eastern diamondbacks are usually very reluctant to defend themselves against humans by biting except as a last resort or when they have been completely surprised and feel threatened. If you see one in the field, you should consider yourself fortunate (many herpetologists have never seen one) and watch it from a safe distance. Never harass or try to capture a diamondback rattlesnake.

before someone is aware of the snake's presence, as in the rare instances when someone steps directly on one in the field.

CONSERVATION Herpetologists consider the diamondback rattlesnake to be one of the most endangered snake species in the country, yet diamondbacks have minimal or no protection in most states, in part because of negative public attitudes and politicians' reluctance to provide protection for a venomous snake. The species does not usually persist in suburban or other developed areas and has apparently been eliminated from Louisiana and from much of its range in other states as well. Whether rattlesnake roundups, some of which still exist in Georgia and Alabama, endanger rattlesnake populations is unresolved, although promoters of such roundups claim that they do not. Some people still kill diamondback rattlesnakes even in wild areas where the snakes belong, and public education appears to be the only way to ensure long-term preservation of this magnificent animal.

Eastern diamondback rattlesnakes have two light stripes on either side of their head.

A Carolina pigmy rattlesnake in ambush posture

Pigmy Rattlesnake *Sistrurus miliarius*

DESCRIPTION Pigmy rattlesnakes are usually gray or grayish brown with a line of dark blotches—and frequently also a gold or orange stripe—running along the middle of the back. They usually have at least one row of smaller dark blotches running along each side, a dark stripe from the eye to the jaw, and two irregular stripes on top of the head. The light-colored belly has irregular dark blotches. Pigmy rattlers have a tiny rattle that is often very difficult to hear. The Carolina pigmy rattlesnake (*S. m. miliarius*) is gray and has a well-defined head pattern. Individuals from Hyde and Beaufort counties in North Carolina have a reddish background color. The head pattern is obscured in the dusky pigmy rattlesnake (*S. m. barbouri*), while the head pattern of the western pigmy rattlesnake (*S. m. streckeri*) is well defined. The rattle segments are fragile and sometimes break off, so some snakes have only a few or no rattles. All three subspecies intergrade with one another in zones of contact (see map).

Some pigmy rattlesnakes have a dark body.

How do you identify a pigmy rattlesnake?

SCALES
Keeled

ANAL PLATE
Single

BODY SHAPE
Heavy bodied

BODY PATTERN AND COLOR
Gray or grayish brown with dark blotches

DISTINCTIVE CHARACTERS
Very tiny rattle; enlarged plates on head

SIZE

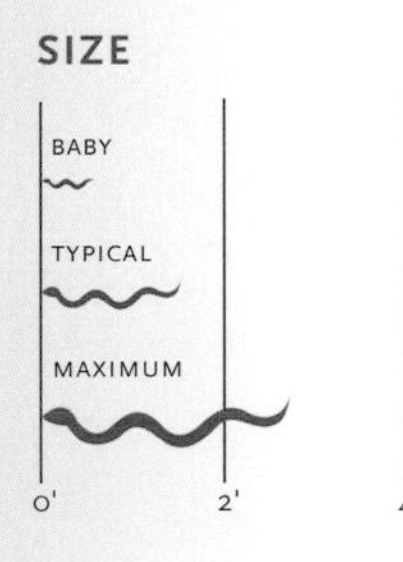

Dusky pigmy rattlesnake from DeLand, Florida

WHAT DO THE BABIES LOOK LIKE? Baby pigmy rattlers look like the adults but have yellow-tipped tails.

OTHER NAMES Pigmy rattlesnakes are known as ground rattlers in many areas of the South.

Did you know?

Pigmy rattlesnakes have proportionally the smallest rattle of any species of rattlesnake.

DISTRIBUTION AND HABITAT Pigmy rattlesnakes are found throughout Florida and in parts of other southeastern states from North Carolina west to Tennessee and south to Louisiana. They vary regionally in their habitat choices, ranging from dry upland sandhills and mixed hardwood and pine forests in some areas to low-lying, sometimes flooded, palmetto stands, floodplains, and marshy habitats in others.

BEHAVIOR AND ACTIVITY Pigmy rattlesnakes hibernate or at least become inactive during the winter in most parts of their geographic range, but even during periods of cold weather some individuals may be found basking on sunny days. Seasonal peaks of activity vary. In many parts of the Coastal Plain, autumn is the period of highest aboveground activity, but pigmy rattlesnakes are also encountered frequently in early to late spring in many areas. Movement overland is more common at night but may also occur during the day. In areas where their habitat becomes flooded, pigmy rattlesnakes will often climb several feet above the ground into bushes, vines, or palmettos.

Pigmy Rattlesnake

Sistrurus miliarius

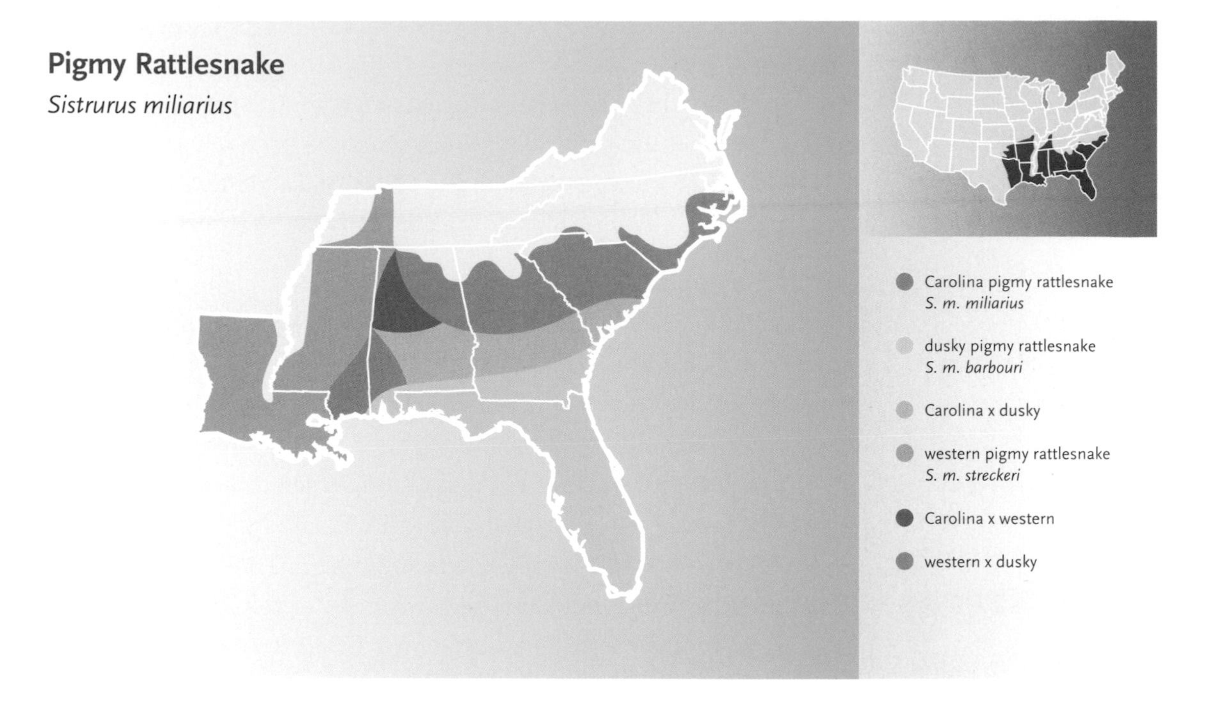

FOOD AND FEEDING Pigmy rattlesnakes feed on a wide variety of small animals including lizards, snakes, frogs, centipedes, and mice. Adults capture prey primarily by ambush, and juveniles may use their yellow-tipped tails to lure prey within striking range. Both adults and juveniles may actively hunt and stalk some prey. Pigmy rattlesnakes use chemical cues when selecting ambush sites. They generally strike and then hold on to their prey until it dies, although larger prey animals such as mice are probably released and then tracked.

Pigmy rattlesnakes from some parts of North Carolina are reddish in color.

The western pigmy rattlesnake found throughout most of Louisiana resembles the eastern subspecies.

REPRODUCTION Pigmy rattlesnakes mate in the fall and spring. Females reproduce every year or every other year. Typically 6 or 7—but occasionally more than 30—babies are born in August or September. Females from the northern parts of the range (e.g., North Carolina) generally give birth to fewer young than those in the south (e.g., Florida). Larger females produce larger and more offspring than do smaller females.

PREDATORS AND DEFENSE Despite being venomous, pigmy rattlesnakes become prey themselves for many other species because of their small size. Any large snake-eating snake (e.g., kingsnakes, racers, and indigo snakes) can eat them, as can large mammals, hawks, and owls. Babies are probably eaten by large frogs, toads, shrews, and larger ground-feeding birds such as turkeys, brown thrashers, and robins. When coiled in pine straw, palmettos, or other natural habitats, these little snakes are exceedingly difficult to see, even by those skilled at finding them. Thus, their response to an encounter with a human is first to remain motionless if in a coiled position, presumably with the assumption of being camouflaged. A pigmy rattlesnake threatened while crawling over the ground will generally coil and face its adversary. In most instances they will not strike unless pestered, and usually bite only if picked up—and sometimes not even then. No venomous snake should ever be picked up, however, because all species have the potential to bite and inject venom.

Pigmy rattlesnakes are active at night in some areas during warm weather.

CONSERVATION Because they are so well camouflaged, pigmy rattlesnakes are probably more common than is apparent in places where they occur. Thus, many populations are probably unknowingly lost to habitat development or to timber management and agricultural techniques that chop or till the soil and remove vegetation. Many pigmy rattlesnakes are killed crossing roads each year.

How dangerous are they? Because of their small size, pigmy rattlesnakes are generally not considered particularly dangerous to humans. Most bites occur when someone picks up a snake or steps on it, and bites in Florida are common. Nearly all victims who receive proper medical treatment recover fully. If you find a pigmy rattlesnake in the field, observe it from a safe distance but do not attempt to capture it or disturb it in any way. If you are concerned because one is found in your yard, have it removed by someone with experience handling venomous snakes.

Coral snakes have a black nose and head followed by a broad yellow ring.

Coral Snake

Micrurus fulvius

DESCRIPTION Coral snakes are slender, smooth-scaled snakes that have red, yellow, and black rings. The red rings always in contact with yellow rings and the black snout distinguish coral snakes from mimics such as the scarlet snake and scarlet kingsnake. The Texas coral snake (*M. f. tener*) has large amounts of black pigment on the red bands, forming spots or blotches, whereas the eastern coral snake (*M. f. fulvius*) has only tiny spots of black on the red bands.

WHAT DO THE BABIES LOOK LIKE? The babies look like miniature adults.

OTHER NAMES The coral snake is sometimes called the harlequin snake or candy cane snake.

DISTRIBUTION AND HABITAT Coral snakes can potentially be found anywhere in Florida but are patchily distributed in the other southeastern states, where they are confined mostly to parts of the Lower Coastal Plain from the Carolinas to Louisiana. They are typically associated with dry habitats such as scrub oak–pine sandhills and, in Florida, hardwood hammocks, but they also thrive in frequently flooded pine flatwoods.

BEHAVIOR AND ACTIVITY Coral snakes may hibernate for several weeks in the northern or inland portions of their geographic range but may be

How do you identify a coral snake?

SCALES
Smooth

ANAL PLATE
Usually divided

BODY SHAPE
Slender

BODY PATTERN AND COLOR
Red, yellow, and black rings

DISTINCTIVE CHARACTERS
Black nose; red and yellow rings touch

SIZE

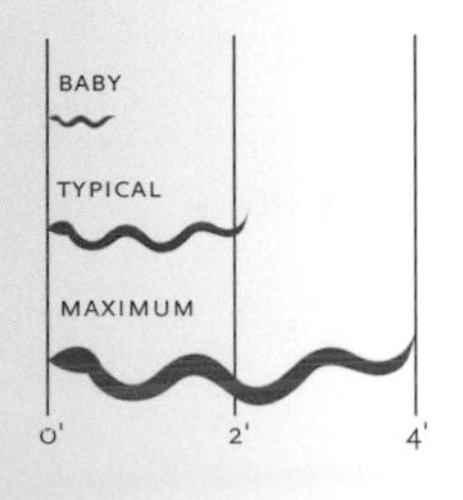

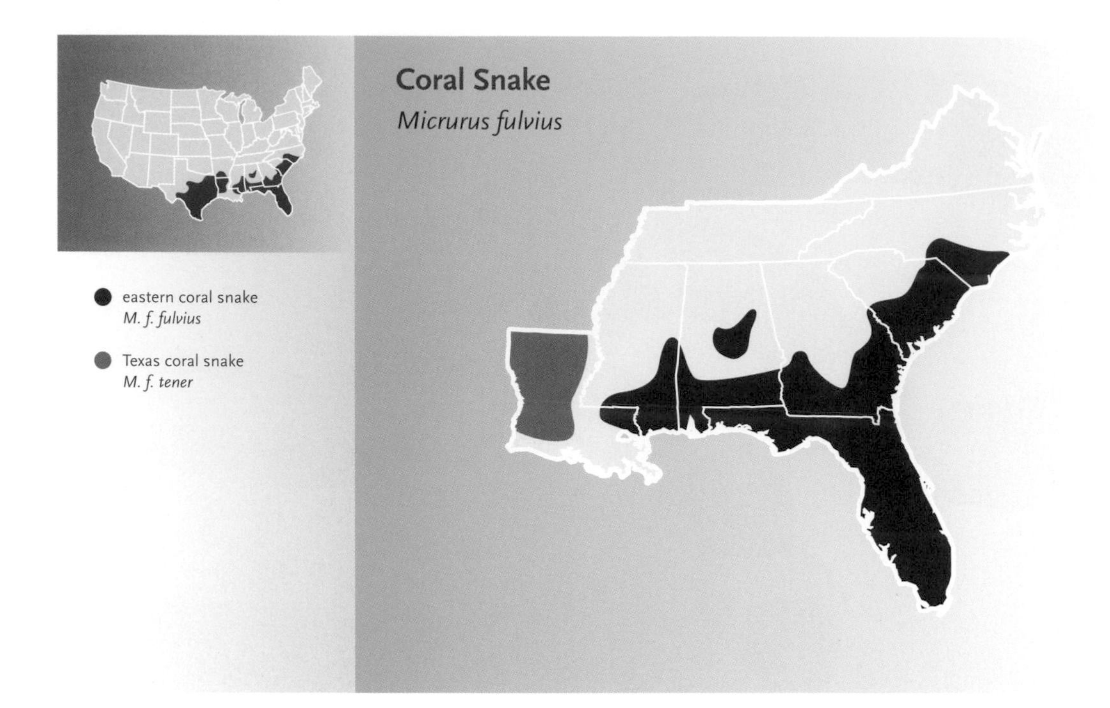

inactive for shorter periods during winter in southern Florida and along the coast. They are characteristically most active in the spring and fall, especially during wet years, and are seen less commonly in the summer. They presumably spend most of their time underground in tunnels left by decayed roots, in burrows of small mammals and insects, and in other natural passageways beneath the surface. Overland travel in the Southeast is almost always during the daytime, and more often during the morning than at other times of the day. Coral snakes seldom if ever climb into trees or bushes.

Coral snakes have a rounded head and round pupils.

FOOD AND FEEDING Coral snakes feed primarily on lizards and other snakes. They are active foragers that search for prey by poking their heads under leaves and other debris and by following underground tunnels created by roots and burrowing animals. Prey are grabbed and chewed until venom is injected and begins to take effect. The venom kills by disrupting the central nervous system. Coral snakes apparently use both visual and chemical cues to detect potential food.

REPRODUCTION Coral snakes mate in the spring and probably also in the early fall. The courting male slowly rubs his head and body over the female until she indicates her readiness to mate by raising her tail. Coral snake eggs are proportionately more elongate than

Scarlet snakes are often mistaken for coral snakes but do not have a black head or broad red rings touching yellow rings, and they have pointed snouts.

Red touch yellow, kill a fellow— red touch black, venom lack

Several nonvenomous snakes that live in the southeastern United States have color patterns similar to the coral snake and are thus considered coral snake mimics. Potential predators might perceive these snakes to be dangerous and thus avoid them.

In the United States you can tell a nonvenomous mimic such as the scarlet kingsnake or the scarlet snake from the venomous coral snake by using the little rhyme above. The venomous coral snake has red bands that touch its yellow bands; the mimics do not.

Important note: This rule does not work in Latin America, where there are many other species of coral snake and many other mimic species.

Scarlet kingsnakes and Louisiana milksnakes resemble coral snakes but differ in that the red rings touch the black rings.

Coral snakes have a black nose and head followed by a broad yellow ring.

those of most other snakes and are laid underground, in leaf litter, or in other moist protected places during the late spring or summer. Clutches usually contain about seven eggs but may total more than a dozen, and the babies hatch in August or September.

PREDATORS AND DEFENSE Coral snakes are preyed on by larger snake-eating snakes such as kingsnakes, indigo snakes, and racers, and even by coral snakes that are larger than themselves. Birds of prey sometimes take coral snakes, but presumably most recognize the distinctive color pattern and avoid them. Coral snakes have been known to kill full-grown hawks imprudent enough to capture them.

A coral snake's first response to danger is to try to escape underground or beneath ground litter. If unable to escape, some individuals will flatten the body, put the head beneath it, and then raise the tail as if it were the head. If a predator goes after the tail, the coral snake might be able to escape or bite the unsuspecting attacker. Coral snakes readily bite if picked up or even approached too closely, but like other venomous snakes will escape to safety if they can. Coral snakes chew on their victims, including people who pick them up, and can sometimes be yanked off before much or any venom is administered.

The broad red rings of coral snakes touch narrow yellow rings.

CONSERVATION Like pigmy rattlesnakes, coral snakes are often present but not seen, and many populations, especially in Florida, have probably been destroyed because of urbanization or other land management practices that destroy upland habitat. In North Carolina, the northernmost part of their range, they are nearly extirpated.

SCIENTIFIC NOMENCLATURE Some herpetologists consider coral snakes from western Louisiana to be a separate species, the Texas coral snake (*M. tener*).

How dangerous are they? Coral snakes have highly toxic venom that attacks the nervous system, including the nerves controlling the breathing muscles. Because of their small size, however, the amount of venom they inject is usually rather small, and a person treated with antivenom is unlikely to die. Bites are very rare and generally occur only when the snake is picked up. If you find a coral snake in the field, it can be safely watched from a short distance away as long as you do not try to pick it up or harass it. If you find a coral snake in your yard or some other area frequented by humans, you should have it relocated to a suitable nearby habitat by someone experienced in handling venomous snakes.

INTRODUCED SPECIES

For centuries, humans have transported animals and plants around the world—sometimes deliberately and sometimes by accident. As a consequence, many species are now found in areas where they do not naturally occur. Most introduced species do not survive. Some, however, adapt particularly well to their new environment and may be extremely successful—more so in some cases than native species. Such introduced species can cause many problems both for humans and for local animal species. Notable examples of introduced species considered environmentally detrimental in the Southeast are kudzu and fire ants.

Many species of reptiles have been introduced into nonnative regions throughout the world. Mediterranean geckos, which have been widely introduced in warm regions of the United States, have caused few or no problems. But other introduced species have been extremely harmful to local species. The brown tree snake (*Boiga irregularis*), native to Indonesia, New Guinea, and northeastern Australia, apparently arrived in Guam in the 1950s aboard a cargo ship. Because they had no natural predators on the island, brown tree snakes rapidly increased and soon caused major damage to populations of native birds and lizards, which were largely defenseless because they had evolved in the absence of snake predators. Brown tree snakes have also caused serious economic problems on Guam because they frequently short out power lines. Special efforts are now being made to prevent brown tree snakes from being introduced into other parts of the world, especially on other islands such as Hawaii.

The brown tree snake native to the Australian and Indonesian regions was unintentionally introduced to Guam, where it has become an invasive species that is a serious threat to other wildlife. The U.S. government maintains strict programs to control accidental shipment of brown tree snakes in cargo, which could prevent the species from being introduced into Hawaii or Miami.

Exotic reptiles have been and continue to be introduced in the southeastern United States, some deliberately, some not. The hospitable tropical climate of southern Florida has allowed several reptile species to become established there. Most of the introduced reptiles are small lizards, such as geckos and anoles, that survive well around humans. The caiman, a medium-sized crocodilian native to Central and South America, has been introduced into aquatic areas around Miami, and several reproducing populations have become established.

Although most of the many nonnative snakes released in the Southeast do not survive, dangerous snakes such as Egyptian cobras and reticulated pythons, presumably released by irresponsible pet owners, have been found. Two species of exotic snakes have apparently established reproducing populations in the Southeast: the Brahminy blind snake (*Ramphotyphlops braminus)* and the Burmese python (*Python molurus bivittatus*).

Brahminy blind snakes from Southeast Asia are now established residents in southern Florida and other parts of the southeastern United States.

Brahminy Blind Snake *Ramphotyphlops braminus*

This native of Southeast Asia was introduced into South Florida, probably in the soil of imported potted plants. It has spread throughout the southern part of Florida and has been found as far north as Newport News, Virginia. Brahminy blind snakes look like shiny black earthworms. They have no enlarged scales on their bellies like most other snakes, and the blunt head and tail are sometimes difficult to tell apart. A small spine on the tip of the tail is used to anchor the snake as it burrows in loose soil. The eyes are extremely small and are visible only as black dots under the head scales. Brahminy blind snakes differ in an unusual way from all native American snakes in that they are all females. Their eggs begin development without being fertilized by sperm. These little snakes have so far caused no noticeable harm. They eat primarily the eggs, larvae, and pupae of ants and termites. Because they are found most often in flower beds, they are sometimes called flowerpot snakes.

Burmese pythons can sometimes exceed 20 feet in length.

Burmese Python *Python molurus bivittatus*

Burmese pythons, a subspecies of the Indian python, have been commonly kept as pets by reptile hobbyists for several decades. These snakes get very large (up to 20 feet long), are very heavy bodied, and are characterized by dark brown blotches on a light brown or golden background. Heat-sensitive pits or holes along the anterior lip scales allow them to sense warm-blooded prey much as pit vipers do. Continuing reports of Burmese pythons both large and small in parts of Everglades National Park suggest that they have established a reproducing population there. Although nonvenomous, large pythons pose a potential danger to domestic pets and humans, especially small children. The impact they are now having or will have in the future on the natural ecosystems of the Everglades is unknown.

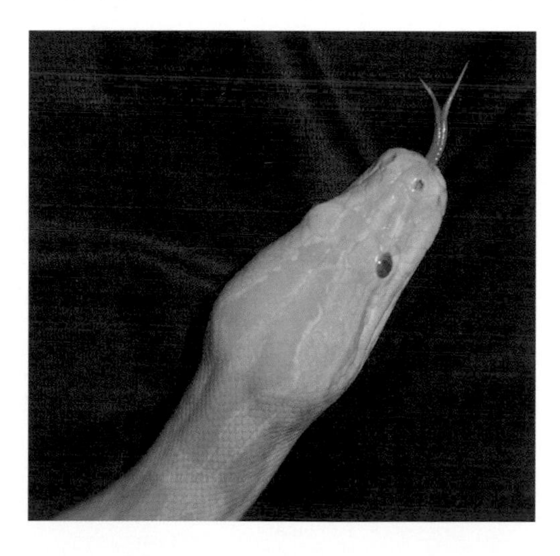

Captive-bred albino Burmese pythons have become popular in the pet trade.

A plastic tube can be used by professional herpetologists to safely restrain venomous snakes during research.

People and Snakes

WHAT IS A HERPETOLOGIST?

Herpetologists are scientists who study reptiles and amphibians (also known as *herpetofauna*). Most herpetologists specialize in a particular discipline or disciplines of biology, such as ecology, physiology, or genetics, and their research often has important implications for other groups of animals in addition to reptiles and amphibians. Many herpetologists today focus on conservation issues related to reptiles and amphibians—for example, investigating the impacts of forestry practices on woodland herpetofauna. Others study the effects of pesticides on reptiles and amphibians, which are particularly sensitive to these chemicals. Countless other research projects involve herpetofauna—so many that some scientific journals are devoted entirely to publishing herpetological research.

Many, perhaps most, professional herpetologists experienced an early taste of their future careers during childhood as they pursued an interest in snakes by collecting them and keeping them as pets. While some herpetologists devote their careers to studying other groups of herpetofauna (frogs, salamanders, lizards, turtles, or crocodilians), the primary appeal for many continues to be snakes. It is a natural extension of an exhilarating journey from youth into adulthood, a journey that for most will last a lifetime.

Did you know?

Herpetologists around the world have described more than 2,800 species of snakes.

Snake traps with funnels are a common technique used by researchers to capture snakes.

Why Do Herpetologists Study Snakes?

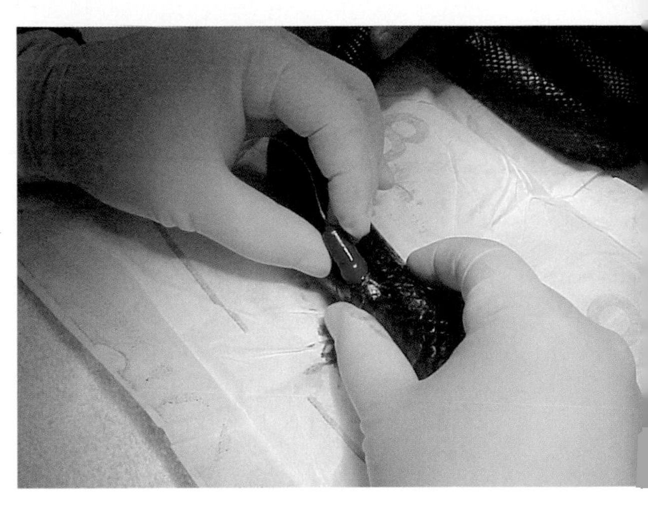

Surgically implanting a transmitter does no harm to the snake if done properly.

Herpetologists study snakes for a variety of reasons. While sheer fascination with a particular species has stimulated many a career in herpetology, that alone does not justify spending considerable money, time, and effort getting to know it better. Some snakes are excellent model organisms for studies of general biological phenomena. For example, research on snakes' metabolism and their ability to fast for long periods might provide insights related to the dietary concerns of domestic animals and humans. In the case of endangered snake species, studies of their basic ecology and behaviors are necessary to understand how to preserve them and their habitats. Other species of snakes are good indicators of environmental quality. Studies evaluating the status of these animals may indicate potential environmental problems.

How Do Herpetologists Study Snakes?

Herpetologists use a variety of techniques to study snakes. Because snakes are often very secretive, capturing them may require special methods. Generally, the best way to catch snakes is to spend a lot of time in their natural habitats. Watersnakes can be grabbed from trees while they are basking over the water, and many snakes can be found crossing roads.

One effective way to capture snakes systematically is to use a drift fence, which is simply a "wall" erected in an area where snakes occur with funnel traps or buckets placed alongside it at intervals. Snakes that encounter the drift fence tend to turn and follow it, and eventually enter or fall into the traps. Another very effective method for capturing snakes is to use large coverboards—usually sheets of plywood or tin—that are placed on the ground and later lifted to see what has taken refuge beneath them.

Captured snakes are usually measured and marked for future identification; some researchers take tissue samples for DNA analysis. Some snakes can be easily marked by clipping small sections from belly scales in identifiable patterns. A more high-tech method involves the use of passive integrated transponders, or *PIT tags*, which are harmlessly injected into the snake's body cavity and then later read by a device similar to a bar code reader in the grocery store. Each PIT tag has a number that individually identifies a particular snake.

Many herpetologists use *radiotelemetry* to study snakes' natural behaviors in the field. A radiotransmitter in the snake emits a signal at a particular

Did you know?

Herpetologists have no reliable way of knowing the exact age of a snake unless its birth date is known.

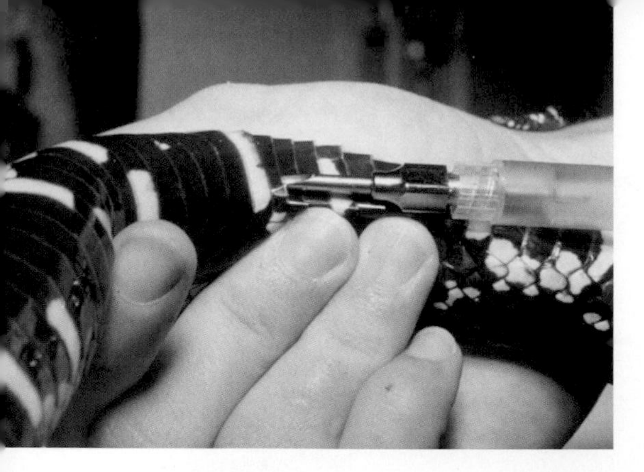

radio frequency that allows a researcher with a receiver and a special antenna to follow it. Some special transmitters even have sensors that allow the researcher to monitor the snake's body temperature. Because a snake cannot wear a radio collar like most other wild animals that scientists study with radiotelemetry, transmitters typically are surgically implanted into the snake's body cavity. This simple surgery causes no harm to the snake if done correctly by an experienced herpetologist.

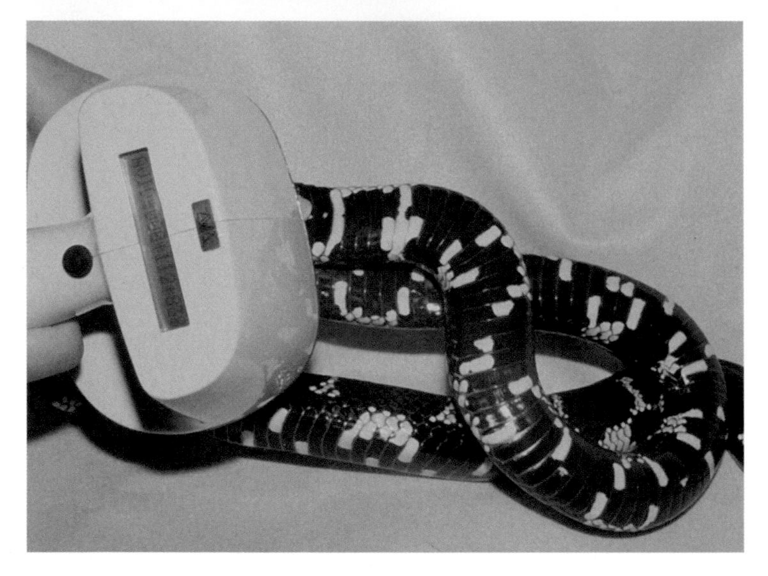

PIT tags are an effective means of permanently identifying individual snakes.

What Rules Must Herpetologists Follow to Study Snakes?

Herpetologists who collect snakes for research or educational purposes, such as during a college class in herpetology, must know and follow the state and federal laws that protect snakes and other wildlife in their area. All southeastern states require anyone who wishes to remove a snake from the wild for research purposes to have a scientific collecting permit, and some states place limits on how many individuals of certain species can be kept at one time. Southeastern snakes protected under the Endangered Species Act (indigo snakes and Atlantic salt marsh snakes) can be handled in the field or kept in captivity *only* by those with a special federal permit. Although learning the laws, which differ from state to state, and obtaining permits may require paperwork and payment of a fee, most rules and regulations are intended for the protection and overall best interest of the animals.

BACKYARD SNAKES

Several species of harmless snakes are often found in areas where people live. These "backyard snakes" offer many opportunities for people to become familiar with a fascinating element of nature and to educate others about them as well. Snakes were on this planet millions of years before humans and houses arrived, so a backyard cannot be viewed as the natural habitat of any snake. However, many species can survive around humans, and some even appear to thrive. At least one southeastern species, the brown snake, is found more often in suburban and urban areas than in its natural habitat in the woods!

An individual of any species can potentially turn up in someone's yard because people have invaded every natural habitat in the Southeast to some degree. Wild animals that have been displaced from their homes may roam in search of food or shelter and may unwittingly enter a newly developed housing area or shopping center. As time goes by, most wildlife disappears, including snakes. But some species persist.

As noted above, the snake species most commonly found in close association with humans across much of the Southeast is the brown snake, although other species may be more prevalent in certain localities. Rat snakes are commonly found in suburban areas, especially if trees and shrubs are present, and corn snakes can also be found in many vegetated suburbs.

Several small, secretive species of snakes do well around houses as long as adequate places to hide are present and their prey base of earthworms, slugs, and insects has not been damaged with pesticides. Ringneck snakes and the earth snakes can turn up under rocks, logs, or other hiding places in a backyard. Although less secretive than the smaller species, the eastern garter snake can show up almost anywhere, too, including in someone's yard or on their porch. Rough green snakes often do quite well in neighbor-

Green snakes can sometimes be found in shrubbery and other vegetation around homes in every southeastern state.

The small brown snake is one of the most common snakes found in many urban areas.

hoods with shrubs, vines, and bushes, but their presence is unpredictable. One of the most unexpected of the backyard snakes is the scarlet snake, sometimes referred to as the "swimming pool snake" in sandhill-area communities because individuals frequently fall into swimming pools during their nighttime activities.

In housing areas or city parks with lakes, northern water snakes and banded water snakes often thrive and become part of the common fauna. Queen snakes have been known to persist in developed areas where streams occur as long as conditions remain unpolluted and crayfish are present.

Less welcome, yet sometimes common, residents of suburban areas are two venomous species—copperheads and pigmy rattlesnakes. When present in an area, both species are probably much more common than the human inhabitants realize because the snakes spend much of their time hiding or inactive. They are seen most often during the spring and fall as they move to and from hibernation sites or in search of mates.

Rat snakes are the most likely species to be found inside a house in many areas of the Southeast because of their climbing ability and relative abundance in many urban and suburban areas. However, any snake can readily enter a house with doors that are at ground level, and domestic cats often bring live snakes and other animals into houses. Once a snake is removed from a house, it is unlikely to be seen again.

Some species of southeastern snakes, especially the large rattlesnakes, almost never persist in developed areas. Unless extensive hiding sites are available, individuals are likely to be discovered and are unlikely to survive most encounters with people. Also, the chances of road mortality are extremely high for such slow-moving snakes.

What do the species of backyard snakes have in common? No general rule can explain why some species persist, some thrive, and others disappear completely from habitats when humans move in. The important point is that some snakes do remain, and their survival encourages us to understand that we can share our surroundings with them.

More and more people are coming to appreciate snakes and would like to find an occasional nonvenomous species around. Homeowners can take several measures to encourage snakes to take up residence in their backyards. Nearly all snakes want spots where they can hide. Boards, pieces of tin, and rock piles are all places where snakes feel safe because they can hide from view. Vegetation that provides hiding places, such as dense shrubbery or tall grass, is likely to attract more snakes than carefully trimmed lawns. Nearby water, such as a pond or small stream, may add to the likelihood of encountering a snake in a yard because almost all aquatic snake species spend some time on land. Although a yard with numerous hiding areas may be "snake friendly," unless snakes are already in the area it may still be devoid of snakes. In addition to hiding places, snakes require different forms of prey. One means to ensure suitable prey for many small snakes is to be cautious about applying pesticides that can accumulate in the animals that snakes eat, such as earthworms, slugs, and insects.

Did you know?

No chemical will effectively repel all snakes without having a major effect on other wildlife and domestic pets as well. Some people claim that sulfur or mothballs will deter snakes. However, you would probably have to use so much sulfur to keep the snakes out that it would keep you from entering the house as well.

Garter snakes are commonly found in residential areas throughout the Southeast.

Some people still prefer to keep all snakes away from their gardens, yards, and houses. The bad news for those people is that there is no effective snake repellant. Mixtures of sulfur and naphthalene (moth balls) are sold by some companies with the promise that they will get rid of snakes, but no product has been proven effective for this purpose except in laboratory tests conducted in enclosed systems. In most situations, the application of enough of any chemical to deter snakes would harm many other animals and would likely also be a nuisance to the human inhabitants. Barriers such as silt fencing or vertical aluminum flashing may effectively deter snakes in some situations, but many snakes are good climbers, and others may crawl under barriers that are not embedded deeply in the soil.

Corn snakes are easy to care for and available from many reptile breeders.

Dealing with venomous snakes around a house or the possibility of their presence can pose a dilemma when small children or outdoor pets are present, even for people who appreciate snakes. Removing a venomous snake from an area inhabited by people is best done by a local snake expert from a university, nature park, or environmental education center. You may also be able to find herpetologists with experience handling venomous snakes at a regional reptile and amphibian society.

SNAKES AS PETS

Many people find snakes fascinating, and snakes are often kept as pets. Captive snakes generally eat infrequently, produce little waste, generally do not carry diseases transmittable to humans, do not require a lot of room, and do not make noise. Therefore, as long as you understand the particular snake's requirements, many are fairly easy and safe to maintain in captivity.

If you are interested in obtaining a snake to keep at home, animals bred in captivity generally make far better pets than wild-caught animals. Many reptile breeders produce a wide variety of high-quality captive-bred snakes that sell for relatively low prices compared with those charged by pet stores. Look for their names on the Internet. The most preferred captive-bred animals are ones born from parents that were also born and raised in captivity. Captive breeding produces snakes that are healthier than many wild-caught specimens and also provides an environmental service by helping to reduce the removal

Kingsnakes are usually hardy, gentle, and attractive, making them among the most popular of snake pets.

A well-constructed cage with a hiding place and water bowl are essential for proper care of a captive snake.

of snakes from the wild. Among species native to the Southeast, several varieties of kingsnakes and corn snakes are very colorful and make excellent pets.

Before purchasing any snake, you should first check local, state, and federal laws that may restrict your ability to legally own exotic or native animals. Some states, for example, have very specific regulations restricting the release of pet snakes into the wild. Then make sure you understand the snake's requirements in captivity. Many books are available on the subject, and much useful information can be found on the Internet as well. Venomous snakes and very large boas or pythons can potentially be dangerous and should not be kept as house pets under most circumstances.

Captive snakes require a suitable cage. Snakes are remarkable escape artists, and they can be very difficult to find once they are loose in most people's houses, so make sure that the cage you choose is secure. Aquaria work well for many species because they are inexpensive and easily cleaned, and lids can be purchased that fit them well. Newspaper, paper towels, or aspen bedding provides a suitable substrate for most species. A water dish filled with clean water must be available at all times. Most snakes will use a "hide box" placed inside the cage and should at least be given a folded newspaper or other material in which they can hide. Snakes must also be kept at appropriate temperatures. Special heaters can be purchased that heat one end of a cage and create a gradient, allowing the snake

Snakes have no equal as escape artists and will take advantage of any opportunity.

to choose from a range of temperatures. Kingsnakes, rat snakes, and many other species will usually feed eagerly on either freshly killed mice available from a pet store or frozen mice that have been thawed.

If you decide you want to keep a snake as a pet, make sure that you are motivated by a sincere interest in the animal and not just fascinated by the idea of keeping something of which many people are afraid. In fact, you should never deliberately scare anyone with a pet snake. Doing so enhances people's negative attitudes toward snakes and toward those who have an interest in their welfare. Additionally, if you are motivated by the "thrill" of keeping a snake, that "thrill" may soon diminish, often resulting in neglect of the animal. Finally, wild-captured snakes should be released only at the site where they were captured, and captive-bred snakes should never be released into the wild.

SNAKE CONSERVATION

Although they are feared or even despised by many people, snakes deserve the same consideration given to any other animal. Snakes are valuable elements of the Southeast's natural ecosystems. They serve both as important prey for many animals and as predators that help control other animal populations. In some areas they are the top predators. Some watersnakes can be found in extremely high densities and thus form a large portion of the animal *biomass* and "stored energy" in the ecosystems in which they live. Watersnakes provide abundant prey for many animals, including birds such as herons and egrets. Rat snakes are also abundant in many habitats, especially around farms, where they play an important role in helping to control rodent populations. In fact, some farmers actually release rat snakes around their barns to eat rats and mice.

Did you know?

The best records of age and longevity are for snakes that have been kept in captivity since birth, especially in zoos where accurate records have been kept.

Snakes can also serve as important *bioindicators* or *biomonitors* of environmental integrity. Their absence or lack of abundance may signal general problems in the environment. For example, many watersnakes cannot survive in unhealthy (e.g., contaminated by pesticides) wetlands, and thus their absence can indicate underlying problems in the environment that are not otherwise obvious. Additionally, because snakes are predators and are high in the food chain, some toxins tend to accumulate in their tissues and are thus more detectable in snakes than in many other organisms.

Finally, as an integral part of natural ecosystems, snakes are no less worthy of our respect and admiration than whales or eagles or other animals on which humans place a high value. Snakes add considerably to the biodiversity of the Southeast, and our encounters with them provide many

Venom is no defense against cars and trucks.

memorable experiences. How many people would ever forget the fascinating spectacle of a rat snake crawling up the trunk of a tree or a watersnake swallowing a catfish?

Threats to Snakes

Unfortunately, like other wild animals, and far more than most, snakes suffer as a result of the activities of humans. Habitat destruction is by far the primary conservation concern for southeastern snakes. Although a few species of snakes may persist around some human habitations, most die out whenever their habitat is altered significantly or destroyed. Human development not only destroys habitat for snakes, but roads, parking lots, and commercial structures separate the remaining pieces of habitat, isolating populations of snakes and making them more prone to extinction. Mortality on roads is a continual threat to snakes during warm months in all parts of the Southeast. And intentional killing of snakes continues in areas where a sense of environmental stewardship has not been instilled in the residents.

Behavioral research can determine how different species of snakes respond to roads.

Although populations of most snake species have declined everywhere throughout the Southeast because of human activities, some species have been hurt more than others. Indigo snakes and southern hognose snakes have disappeared from much of their previous ranges as a result of human activities. The eastern diamondback rattlesnake, the largest species of rattlesnake in the world, is now apparently extinct in Louisiana, and only remnants of populations persist in North Carolina. Likewise, southern hognose snakes can no longer be found in Alabama or Mississippi. Clearly, the decline of snake populations in much of the Southeast reflects the situation of many wildlife species whose survival is not compatible with widespread habitat loss and degradation.

Habitat destruction resulting from unconstrained development is the greatest environmental threat to most species of snakes.

Conservation Laws

Persuading legislators to pass laws protecting snakes, especially venomous species, is a difficult task. Laws and regulations designed to protect certain species from commercial collection or wanton killing have been implemented in some states. However, few if any laws effectively prevent habitat from being destroyed solely because an area serves as an important environment for snakes. The indigo snake and Atlantic salt marsh snake are among the very few snakes in the entire country to receive any federal protection from the Endangered Species Act. But laws and regulations are only as effective as those who enforce them. Some law officers and judges remain among those who have an irrational fear and dislike of snakes, and do not uphold the laws designed to protect them.

Did you know?

The Southeast is home to two species of snakes (indigo snake and Atlantic salt marsh snake) that are on the federal endangered species list.

Snakes' Best Hope

The best hope to protect snakes' habitats and prevent malicious killing lies in public education. People who understand the importance of snakes as natural parts of our ecosystems are more likely to support laws to protect them. It is important to remember that habitat loss is not a threat to snakes alone but is also the primary threat to other wildlife. Snakes will benefit from habitat protection designed for other animals, and vice versa. Once

Black swamp snakes can be abundant in a localized habitat while suffering serious population losses in a many parts of their geographic range.

Educating children is one way to create positive public attitudes about reptile conservation.

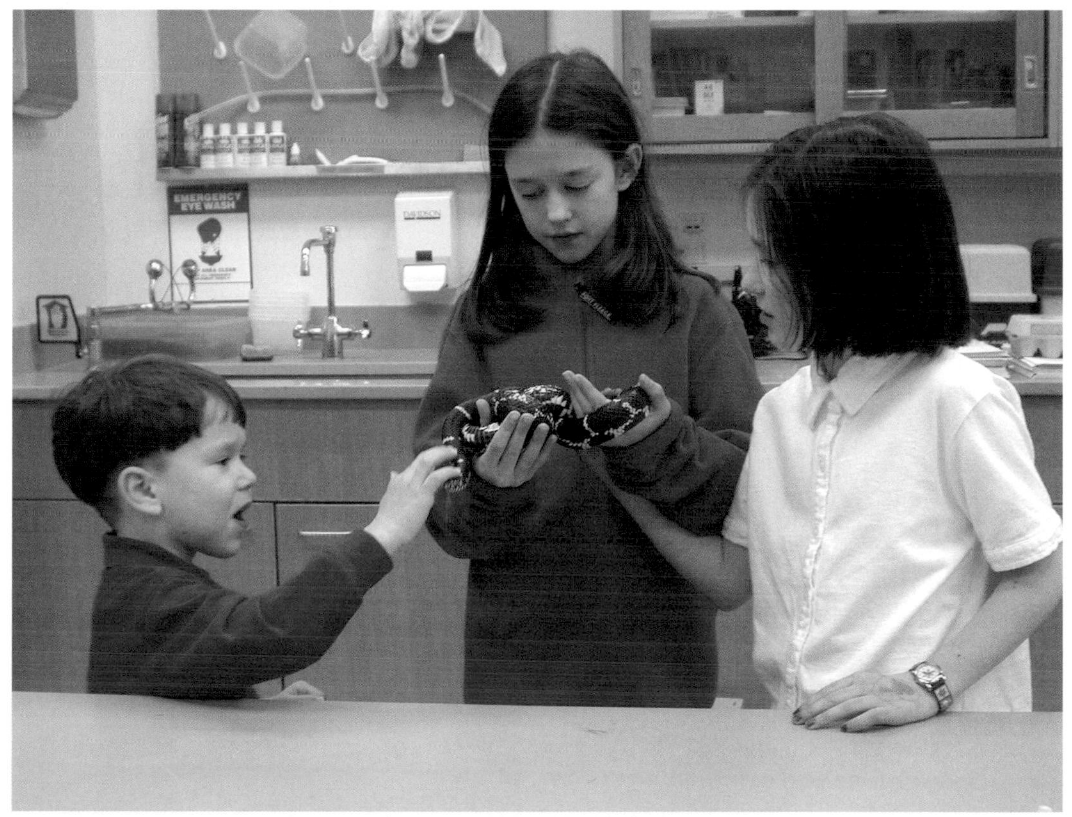

Because they are so small,
pigmy rattlesnakes are
generally not considered life-
threatening to humans.

An attractive appearance can help
to convince some people that
snakes are not necessarily bad.

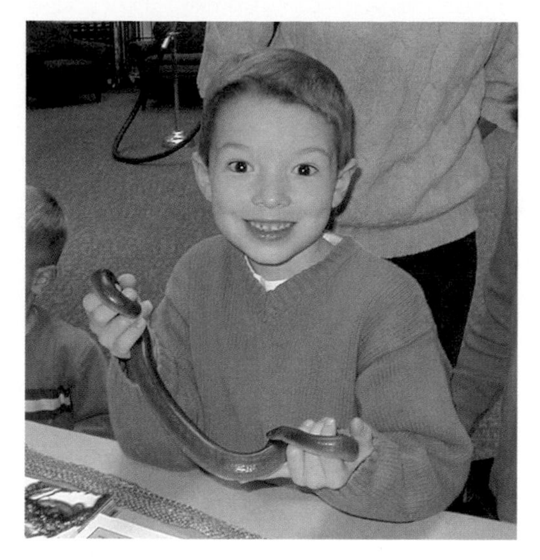

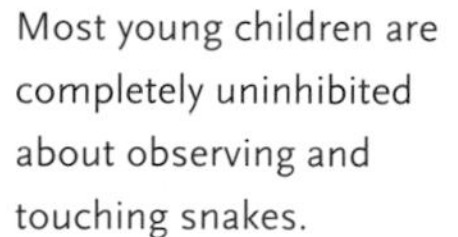

Most young children are completely uninhibited about observing and touching snakes.

the general public learns to appreciate snakes as valued components of our natural world, laws and attitudes will change, and we will be assured of sharing the Southeast with the snakes and other creatures that are its rightful inhabitants for many, many years to come.

ATTITUDES ABOUT SNAKES

People's feelings about snakes tend to be strongly polarized. They may love them or hate them, but very few people are indifferent to them. It is difficult to explain humans' powerful love/hate relationship with snakes. One theory says that humans have an innate fear of snakes (ophidiophobia) that is perpetuated by society because a few species are potentially dangerous. The fact that many people today have an appreciation for snakes rather than an aversion clearly demonstrates that the fear can be overridden.

Familiarity is the best way for most individuals to transform their fear of snakes into respect and even admiration. Educating the public about the fascinating behavior, ecological value, and minimal threat associated with southeastern snakes is the first step toward developing a general attitude that snakes have far more to offer than most people realize.

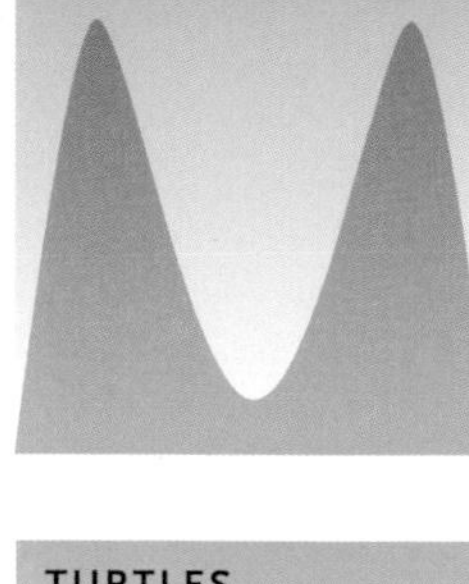

TURTLES

Proportion of Population

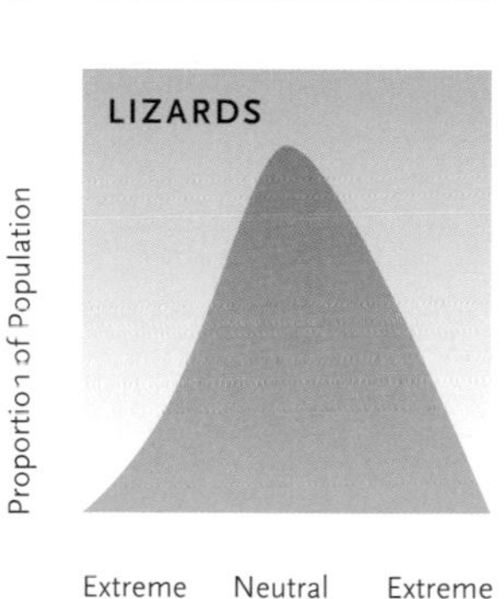

While some people despise snakes and others find them fascinating, few people have neutral feelings about these animals, compared to other reptiles such as turtles and lizards. These charts reflect the authors' perceptions based on their more than 60 years of combined experience communicating with the public about snakes and other reptiles.

What kinds of snakes are found in your state?

Occurrence by state of the 53 species of snakes native to the Southeast, listed in the order in which they appear in the book

COMMON NAME	LA	MS	AL	GA	FL	SC	NC	VA	TN
Smooth earth snake	●	●	●	●	●	●	●	●	●
Rough earth snake	●	●	●	●	●	●	●	●	●
Brown snake	●	●	●	●	●	●	●	●	●
Red-bellied snake	●	●	●	●	●	●	●	●	●
Southeastern crowned snake	●	●	●	●	●	●	●	●	●
Florida crowned snake				●	●				
Rim rock crowned snake					●				
Flat-headed snake	●								
Short-tailed snake					●				
Pine woods snake	●	●	●	●	●	●	●		
Eastern worm snake	●	●	●	●		●	●	●	●
Western worm snake	●								
Ringneck snake	●	●	●	●	●	●	●	●	●
Scarlet snake	●	●	●	●	●	●	●	●	●
Rough green snake	●	●	●	●	●	●	●	●	●
Smooth green snake								●	
Eastern garter snake	●	●	●	●	●	●	●	●	●
Eastern ribbon snake	●	●	●	●	●	●	●	●	●
Western ribbon snake	●	●							●
Eastern hognose snake	●	●	●	●	●	●	●	●	●
Southern hognose snake		●	●	●	●	●	●		
Mole and prairie kingsnakes	●	●	●	●	●	●	●	●	●
Scarlet kingsnake and milksnake	●	●	●	●	●	●	●	●	●
Common kingsnake	●	●	●	●	●	●	●	●	●
Pine snake	●	●	●	●	●	●	●	●	●
Louisiana pine snake	●								
Rat snake	●	●	●	●	●	●	●	●	●
Corn snake	●	●	●	●	●	●	●	●	●
Racer	●	●	●	●	●	●	●	●	●
Coachwhip	●	●	●	●	●	●	●		●
Eastern indigo snake		●	●	●	●				
Black swamp snake			●	●	●	●	●		
Striped crayfish snake				●	●				
Glossy crayfish snake	●	●	●	●	●	●	●	●	
Graham's crayfish snake	●	●							
Queen snake		●	●	●		●	●	●	●
Northern watersnake	●	●	●	●	●	●	●	●	●
Southern banded watersnake	●	●	●	●	●	●	●		●
Salt marsh snake	●	●	●		●				
Plain-bellied watersnake	●	●	●	●	●	●	●	●	●
Diamondback watersnake	●	●	●						●
Brown watersnake			●	●	●	●	●	●	
Eastern green watersnake				●	●	●			
Western green watersnake	●	●	●		●				●
Mud snake	●	●	●	●	●	●	●	●	●
Rainbow snake	●	●	●	●	●	●	●	●	
Copperhead	●	●	●	●	●	●	●	●	●
Cottonmouth	●	●	●	●	●	●	●	●	●
Timber/canebrake rattlesnake	●	●	●	●	●	●	●	●	●
Eastern diamondback rattlesnake	●	●	●	●	●	●	●		
Pigmy rattlesnake	●	●	●	●	●	●	●		●
Coral snake	●	●	●	●	●	●	●		
TOTAL	41	41	41	41	43	38	37	30	32

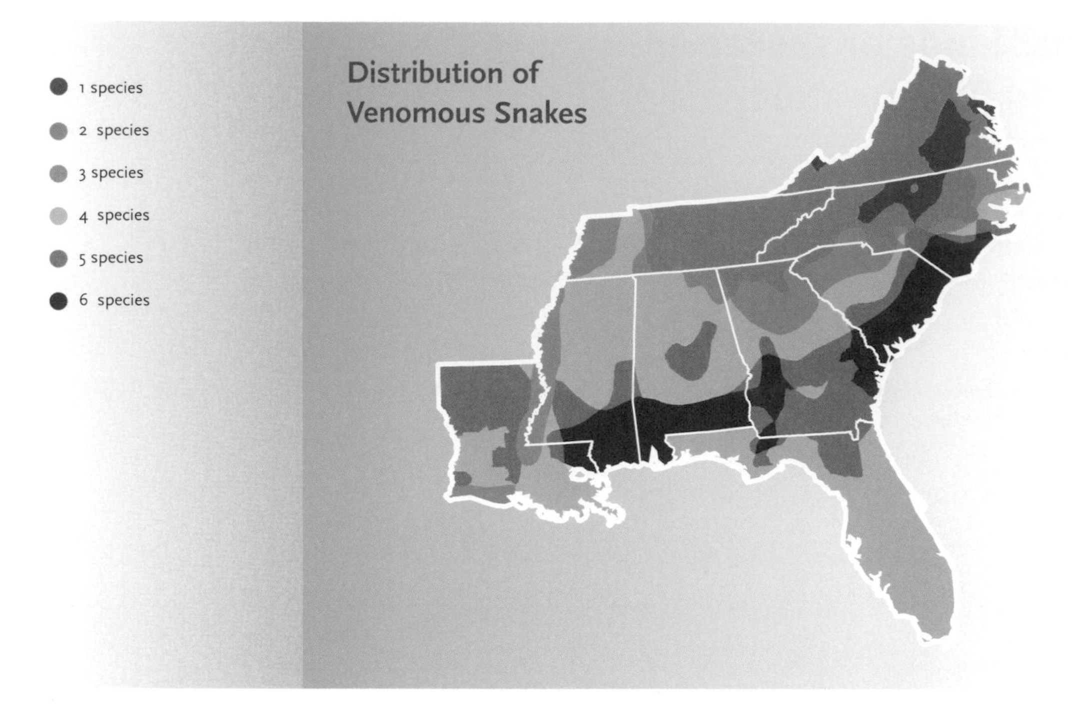

At least one native species of venomous snake is found throughout every southeastern state. Some regions have as many as six venomous species (e.g., southern Alabama and coastal South Carolina) whereas other regions may have only one (e.g., parts of North Carolina and Virginia). The color coding on the map indicates how many species are found in different areas.

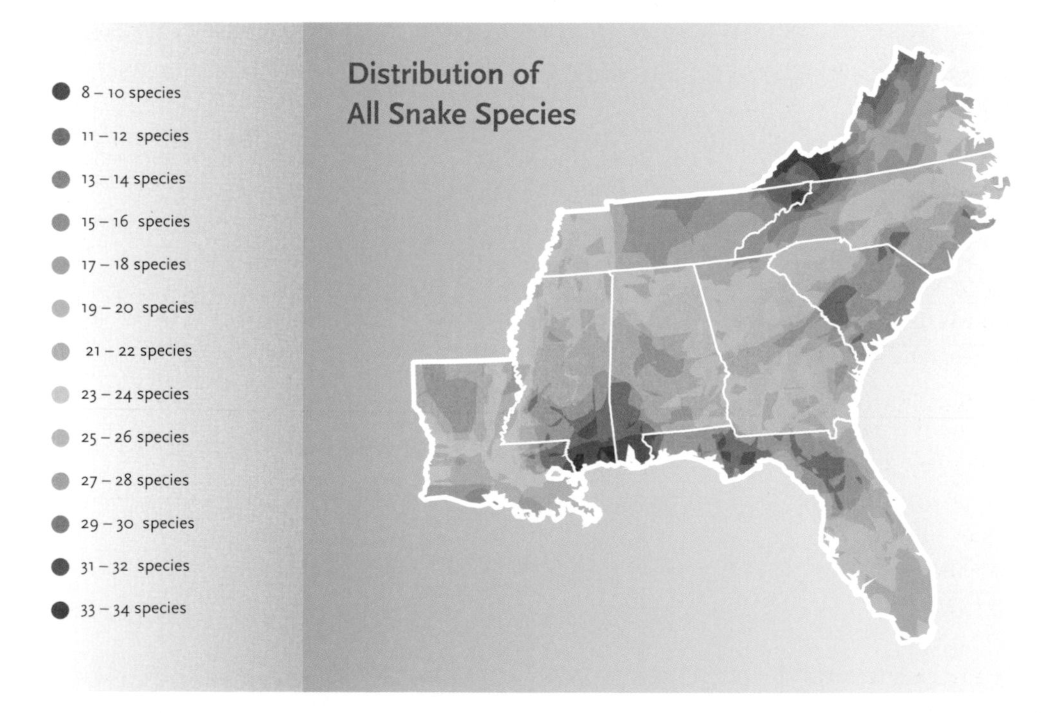

The number of native species of snakes in an area varies greatly among regions in the Southeast, with parts of southern Mississippi having the highest numbers with more than 30 species. The color coding on the map indicates approximately how many species are found in different areas. For example, the central portion of northern Louisiana and central panhandle of Florida have 26–29 species compared to southeastern Tennessee and northern Georgia with 17–20 species.

Glossary

Aestivation A period of inactivity during dry and/or hot periods in which animals wait for conditions suitable for feeding and other activities to improve.

Albino An animal completely lacking pigment that provides the color to skin and eyes. Animals lacking only dark pigment, or melanin, are often referred to as "albino" but should more properly be referred to as "amelanistic."

Amphiuma Any of three species of large, eel-like, aquatic salamanders inhabiting the southeastern United States. They have tiny legs and eyes and prefer swampy habitats with aquatic vegetation.

Anal plate The last scale on the belly of a snake, which covers the cloaca and precedes the scales covering the undersurface of the tail.

Anterior Referring to the end of the animal toward the head.

Antivenom A substance administered to a snakebite victim to serve as an antidote or neutralizer of venom injected by a snake. Antivenoms are produced by injecting horses, rabbits, or other animals with venom and then removing their blood to produce a serum. The antivenom commonly used for bites of pit vipers differs from that used for coral snake bites. Also referred to as "antivenin."

Biodiversity Referring to the numbers, distribution, and abundance of species within a given area.

Bioindicator A species whose health or condition, either at the individual level or at the population level, is indicative of the condition of the habitat or ecosystem as a whole.

Biomass The weight of living things in the environment.

Biomonitor To measure and record the number, types, and characteristics of organisms.

Brumation A period of inactivity of "cold-blooded" animals, or ectotherms, during cold periods. *See also* Hibernation.

Cloaca A single opening that serves as the passageway to the outside of an animal for the urinary, digestive, and reproductive tracts.

Clutch A group of eggs laid together at one time by a single individual.

Cold-blooded A nontechnical term that refers to animals whose body temperature is largely determined by environmental conditions and the thermoregulatory behavior of the animal.

Concertina locomotion A method of locomotion used by snakes in which they anchor the posterior portion of their body, extend the anterior portion, and then pull themselves forward.

Courtship The process that precedes mating in snakes and usually involves actions such as rubbing or biting by the male to increase the receptivity of a female to mating.

Diurnal Active during the daytime.

Dorsal Referring to the back of an animal.

Ecdysis The process of shedding the outer layer of skin.

Ecology The study of how organisms interact with their environment.

Ectotherm An animal whose body temperature is largely determined by environmental conditions and the thermoregulatory behavior of the animal.

Endangered Referring to a species or population that it is considered at risk of becoming extinct.

Endemic Referring to a species found only in a particular geographic location and nowhere else.

Endotherm An animal that maintains a high body temperature primarily through the use of heat generated by a high metabolic rate.

Extinct Referring to species with no living individuals.

Extirpated Referring to the elimination of a species from a particular region; local extinction.

Family A taxonomic group containing one genus or two or more closely related genera.

Flatwoods Habitat of the lower Coastal Plain of the southern Atlantic and Gulf regions characterized by pines and isolated wetlands in low-lying areas.

Generalist An animal that does not specialize on any particular type of prey or is not restricted to a particular habitat.

Genus (pl. genera) A taxonomic grouping of one species or more than one closely related species.

Hemipenis (pl. hemipenes) One of two penises possessed by individuals of the order Squamata, which includes snakes, lizards, and amphisbaenians.

Hemotoxic Referring to venom that acts by destroying tissues.

Herpetofauna The species of amphibians and reptiles that inhabit a given area.

Herpetologist A scientist who studies snakes, other reptiles, and amphibians.

Hibernation A period of inactivity during cold periods. Also known as "brumation" in reptiles.

Hybrid (v. hybridize) An intermediate form resulting from mating and genetic mixing between individuals of two species.

Incubation period The time period between when eggs are laid and when they hatch.

Intergrade An intermediate form of a species resulting from mating and genetic mixing between individuals of two or more subspecies within a zone where their ranges overlap. Intergrade specimens may possess traits of all subspecies involved.

Jacobson's organs Organs in the roof of the mouth of snakes (and lizards) that detect chemicals from the animal's environment.

Keeled scale A snake scale that has a ridge down the center running parallel to the snake's body. Also known as "rough" scale.

Lateral undulation A method of locomotion used by snakes in which they use waves of the body to push against the ground's surface and move the body forward.

Mimicry A condition in which an animal looks or acts like something else, often a more dangerous animal.

Neurotoxic Referring to venom that acts by disrupting the proper functioning of the nervous system.

Nocturnal Active at night.

Pheromone A chemical that is released by an animal and used as a signal to other animals of the same species. Female snakes will often release pheromones to attract male snakes during the mating season.

Phylogeny The evolutionary relationships among different groups and species of animals.

PIT tag A Passive Integrated Transponder, a glass-encapsulated electronic device injected into the body cavity of animals for identification purposes. PIT tags emit signals that can be read by a PIT tag reader in close proximity.

Pit viper A venomous snake belonging to the family Viperidae and having a heat-sensitive pit between the eye and nostril. The family Viperidae includes the cottonmouth, the copperhead, and rattlesnakes.

Posterior Away from the head of an animal and toward the tail.

Radiotelemetry A method using a radiotransmitter attached to or implanted in an animal to track its movements by locating it using a directional antenna and radio receiver.

Rear-fanged Referring to the characteristic of some snakes of having enlarged teeth in the back of the mouth.

Rectilinear locomotion A method of locomotion used by snakes in which they crawl in a straight line and use their ribs and belly scales to push themselves forward. Rectilinear locomotion is generally used by large, slow-moving snakes when they are not in a hurry.

Rookery A site associated with the development or care of the young.

Sandhills Habitat in the southeastern Coastal Plain characterized by sandy soils, rolling topography, scrub oak, and longleaf or slash pine.

Sidewinding A method of locomotion used on unstable substrates such as sand by some snakes in which they lift the body off of the ground to move forward, leaving a series of unconnected tracks that are parallel to each other but at almost right angles to the direction of movement.

Smooth scale A snake scale that is completely flat with no ridge down the center.

Specialist An animal restricted in its choice of diet or habitat.

Species Typically an identifiable and distinct group of organisms in which individuals are capable of interbreeding and producing viable offspring under natural conditions.

Subspecies A taxonomic unit or "race" within a species, usually defined as morphologically distinct and occupying a geographic range that does not overlap with that of other "races" of the species. Subspecies may interbreed naturally in areas of geographic contact (*see* Intergrade).

Taxonomy The scientific field of classification and naming of organisms.

Ventral Referring to the belly or underside of an animal.

Warm-blooded A nontechnical term that refers to an animal that maintains its body temperature primarily through the use of metabolic heat.

Further Reading

Ashton, R. E. Jr., and P. S. Ashton. 1981. Handbook of Reptiles and Amphibians of Florida. Part 1: The Snakes. Windward Publishing, Miami, Fla.

Bartlett, Richard D., and Patricia Bartlett. 2003. Florida's Snakes: A Guide to Their Identification and Habits. University of Florida Press, Gainesville.

Behler, J. L., and F. W. King. 1979. The Audubon Society Field Guide to North American Reptiles and Amphibians. Alfred A. Knopf, New York.

Campbell, Jonathan A., and William W. Lamar. 2004. The Venomous Reptiles of the Western Hemisphere. 2 vols. Comstock Books in Herpetology. Cornell University Press, Ithaca, N.Y.

Carmichael, P., and W. Williams. 2001. Florida's Fabulous Reptiles and Amphibians. World Publications, Tampa, Fla.

Conant, R., and J. T. Collins. 1998. A Field Guide to Reptiles and Amphibians of Eastern and Central North America. Third expanded edition. Houghton Mifflin, Boston.

Dorcas, M. E. 2004. A Guide to the Snakes of North Carolina. Davidson College Herpetology Laboratory, Davidson, N.C.

Dundee, H. A., and D. A. Rossman. 1989. The Amphibians and Reptiles of Louisiana. Louisiana State University Press, Baton Rouge.

Ernst, C. H., and E. M. Ernst. 2003. Snakes of United States and Canada. Smithsonian Books, Washington, D.C.

Ernst, C. H., and G. R. Zug. 1996. Snakes in Question. Smithsonian Institution Press, Washington, D.C.

Gibbons, Whit. 1983. Their Blood Runs Cold: Adventures with Reptiles and Amphibians. University of Alabama Press, Tuscaloosa.

Gibbons, J. Whitfield, and Michael E. Dorcas. 2004. North American Watersnakes: A Natural History. University of Oklahoma Press, Norman.

Gibbons, Whit, and Patricia J. West, eds. 1998. Snakes of Georgia and South Carolina. Savannah River Ecology Laboratory Herp Outreach Publication 1, Aiken, S.C.

Greene, H. W. 1997. Snakes: The Evolution of Mystery in Nature. University of California Press, Berkeley.

Jackson, J. J. 1983. Snakes of the Southeastern United States. Cooperative Extension Service. University of Georgia, Athens.

Lohoefener, R., and R. Altig. 1983. Mississippi Herpetology. Mississippi State University Research Center, NSTL Station, Miss.

Martof, B. S., W. M. Palmer, J. R. Bailey, J. R. Harrison III, and J. Dermid. 1980. Amphibians and Reptiles of the Carolinas and Virginia. University of North Carolina Press, Chapel Hill.

Mitchell, J. C. 1994. The Reptiles of Virginia. Smithsonian Institution Press, Washington, D.C.

Mount, R. H. 1975. The Reptiles and Amphibians of Alabama. Auburn University Agricultural Experiment Station, Auburn, Ala.

Palmer, W. M., and A. L. Braswell. 1995. Reptiles of North Carolina. University of North Carolina Press, Chapel Hill.

Pinder, M. J., and J. C. Mitchell. 2001. A Guide to the Snakes of Virginia. Wildlife Diversity Special Publication 2. Virginia Department of Game and Inland Fisheries, Richmond.

Rossman, Douglas A., Neil B. Ford, and Richard A. Seigel. 1996. The Garter Snakes: Evolution and Ecology. Animal Natural History Series, vol. 2. University of Oklahoma Press, Norman.

Tennant, A., and R. D. Bartlett. 2000. Snakes of North America: Eastern and Central Regions. Gulf Publishing Company, Houston, Tex.

Zim, H. S., and H. M. Smith. 2001. Reptiles and Amphibians. A Golden Guide. St. Martin's Press, New York.

Acknowledgments

We appreciate the support and understanding of our families during the preparation of this book: Carolyn, Laura, Michael, Jennifer, Allison, and Parker Gibbons; Susan Lane and Keith Harris; Jennifer and Jim High; and Tammy, Taylor, Jessika, and Zachary Dorcas.

We owe special thanks to two people, Teresa Carroll and Margaret Wead, without whose help we would have been many more months completing this book. Their constant attention to details in the preparation of photographic slides and digital images, copying text and drawings, and filing materials was invaluable.

We appreciate constructive comments on individual species accounts and help with organizational aspects from herpetology students and staff at the University of Georgia, Savannah River Ecology Laboratory, and Davidson College, particularly Kimberly M. Andrews, Erin Clark, Luke A. Fedewa, Xavier Glaudas, Gabrielle Graeter, Judy Greene, Pierson Hill, Natalie L. Hyslop, Brian Metts, Mark S. Mills, Tony Mills, Sean Poppy, Steven Price, Brian D. Todd, Ria Tsaliagos, Tracey D. Tuberville, Lucas Wilkinson, J. D. Willson, Christopher T. Winne, and Cameron A. Young. We owe special thanks to J. D. Willson, Cameron A. Young, Kimberly M. Andrews, Susan G. Harris, and Steven J. Price for reading and providing useful comments on the entire book.

Joe Mitchell of the University of Richmond and John Jensen of the Georgia Department of Natural Resources provided helpful suggestions, information, and guidance throughout the book's preparation. We appreciate the help of Paul Moler of the Florida Fish and Wildlife Conservation Commission for providing insights into the life and habits of the rare short-tailed snake.

We are very grateful to the many herpetologists throughout the country who provided us with a spectacular array of snake slides and digital images from which to select. The magnificent color images generously offered by so many experts in herpetological photography have added immensely to the value of this book as a practical field guide for identification of southeastern snakes. The following individuals provided material for us to examine: Richard D. Bartlett, Jeff Beane, Ted Borg, Kevin Enge, John Hall, Terry Hibbitts, Toby Hibbitts, Troy Hibbitts, Pierson Hill, RL Hodnett, John B. Iverson, John Jensen, Bradley Johnston, Barry Mansell, Peter May, Brad Moon, Robert N. Reed, Drew Sanders, David E. Scott, John Sealy, Eric Stine, R. Wayne Van Devender, James Van Dyke, Laurie J. Vitt, John White, Lucas Wilkinson, J. D. Willson, Chris Winne, and Cameron A. Young. We thank Leslie Rissler, Deno Karapatakis, Steve Harper, Bill Ringle, and Pete Rothfus for assistance with the range maps.

Finally, because this book supports the efforts of Partners in Amphibian and Reptile Conservation (PARC) to promote education about reptiles and amphibians, we thank the many PARC members who offered encouragement, advice, and enthusiastic support.

Credits

The authors would like to thank the following individuals and organizations for providing photographs:

Richard D. Bartlett
Photographs on pages i, ii, vii, 6–7 (bottom), 8 (bottom), 11 (bottom), 14 (top), 18 (left), 19 (bottom), 25 (bottom), 36–37, 45 (top), 46 (top and left), 47 (bottom), 52, 53, 54 (bottom), 55, 63, 66 (middle), 68–69, 73 (middle), 74, 81 (bottom), 82–83 (bottom), 83 (middle), 85 (bottom), 91, 92 (top), 94, 95 (bottom), 99 (middle), 100–101, 104 (swatches 1, 6), 105 (bottom), 108 (both), 109, 112–113 (bottom), 113 (swatches 2, 3, 7, and 8), 114 (bottom), 116, 117 (bottom), 119 (top), 120 (top), 122 (top), 122–123 (bottom), 123 (swatches 2, 3, 6, and 7), 124 (bottom), 125 (bottom), 129, 151 (bottom), 152 (top), 157, 159 (right and bottom), 161 (bottom), 162, 163 (bottom), 164, 171, 172 (bottom), 173, 175 (bottom), 176–177 (bottom), 185 (top), 197, 202, 203, 209 (right), 211, 213 (top right), 214, 218.

Ted Borg
Photographs on pages 185 (bottom), 212 (left), 231 (top).

Coordinating Group on Alien Pests
Photograph on page 217.

Adam Dawson
Photograph on page 188 (bottom).

Eugene Dorcas
Photograph on page 223.

Mike Dorcas
Photographs on pages viii, 30 (both), 186, 188 (top and middle), 233 (top and bottom), 235.

Kevin Enge
Photographs on pages v, 57, 66–67, 80 (top), 103 (bottom), 104 (swatch 5), 106 (bottom), 148 (bottom), 170, 193 (top), 225.

Whit Gibbons
Photographs on pages 2 (bottom), 15 (top left), 21, 220, 222 (both), 231 (bottom).

T. Hibbitts
Photographs on pages 104 (swatch 4), 113 (swatch 1), and 123 (swatch 1).

Pierson Hill
Photographs on pages 1 (bottom), 87 (top), 89 (bottom).

RL Hodnett
Photograph on page 219 (bottom).

John B. Iverson
Photographs on pages 6 (bottom left), 16 (top), 58, 60–61, 71 (bottom), 98–99 (bottom), 104 (swatch 3), 144 (bottom), 180 (bottom), 212–213 (bottom).

John Jensen
Photograph on page 43 (top).

Barry Mansell
Photograph on page 138.

Peter May
Photograph on page 139.

Brad Moon
Photographs on pages 1 (top right), 14 (bottom), 18 (right), 141, 143, 144 (top), 178 (top), 179, 191.

Robert N. Reed
Photographs on pages 75 (bottom) and 226.

Drew Sanders
Photograph on page 204.

Savannah River Ecology Laboratory
Photograph on page 22.

David E. Scott
Photographs on pages 7 (top), 12 (left and bottom), 13 (right), 23 (top), 26, 32, 80 (bottom), 89 (top), 90, 113 (swatch 6), 115 (bottom), 131, 132, 161 (top), 163 (top), 169, 172 (top), 194, 224 (both), 228–229 (bottom).

John Sealy
Photograph on page 9 (top).

Eric Stine
Photograph on page 206 (top).

R. Wayne Van Devender
Photographs on pages 4 (bottom), 5 (both), 6 (top and middle),

R. Wayne Van Devender (cont.)
8 (top), 9 (bottom), 10 (second, third, and bottom), 13 (left), 15 (top right), 16 (bottom), 19 (top right and bottom right), 23 (middle), 24, 28, 38–39 (bottom), 40 (both), 42, 43 (bottom), 44, 45 (bottom), 46 (bottom right), 47 (top right), 48 (both), 49, 50, 51 (both), 56 (both), 59, 62, 64 (both), 67 (right top and bottom), 73 (bottom), 82 (top), 87 (right), 88 (bottom), 92 (bottom), 93, 96–97, 98 (middle), 99 (top), 103 (top), 104 (swatch 2), 105 (top), 106 (top), 107, 110, 112 (top), 113 (swatch 5), 114 (top), 119 (bottom), 120 (bottom), 123 (swatches 4 and 8), 123 (right), 124 (top), 125 (top), 137 (bottom), 142 (top), 146 (left), 146–147, 148 (top), 154 (both), 156 (right top and bottom), 158, 159 (top), 160 (bottom), 163 (middle), 165, 166, 167, 168, 174 (bottom), 175 (top), 177 (right), 181, 187, 190, 192, 193 (bottom), 196, 198 (top), 201, 209 (left), 210, 215, 219 (top), 227.

James Van Dyke
Photographs on pages 15 (bottom right), 54 (top right), 70–71, 113 (swatch 4), 115 (top), 189, 231 (middle).

Lucas R. Wilkinson
Photographs on pages 12 (top right), 118, 221.

J. D. Willson
Photographs on pages iii, 1 (top left), 2 (top), 3, 4 (top), 10 (top), 10–11 (top), 15 (bottom left), 17, 20 (top), 23 (bottom), 25 (top), 29 (all), 38 (top), 60 (top), 65, 66 (top), 73 (top), 75 (top left), 76, 77, 78 (both), 79, 81 (top), 83 (top right), 85 (top right), 86, 95 (top), 97 (middle and bottom), 111 (both), 117 (top), 121 (both), 123 (swatch 5), 126 (both), 127, 128, 133, 135, 136, 137 (top), 140, 142 (bottom), 149, 150, 151 (top), 152 (middle and bottom), 153, 155, 156 (left), 160 (top), 176 (top and middle), 180 (top), 182, 183, 184, 198 (bottom), 200, 206–207 (bottom), 208, 213 (top left), 220, 228 (top), 232, 234 (all).

Chris Winne
Photographs on pages 20 (bottom), 88 (top), 147 (right), 229 (top and middle).

Cameron A. Young
Photograph on page 178 (bottom).

Index of Scientific Names

Boldface page numbers refer to species accounts.

Index of Common Names

Boldface page numbers refer to species accounts.